Chemistry of Food Packaging

Chemistry of Food Packaging

Charles M. Swalm,
Editor

A symposium sponsored by
the Division of Agricultural
and Food Chemistry at the
166th Meeting of the
American Chemical Society,
Chicago, Ill., Aug. 30, 1973.

ADVANCES IN CHEMISTRY SERIES **135**

AMERICAN CHEMICAL SOCIETY

WASHINGTON, D. C.   1974

Library of Congress CIP Data

Chemistry of food packaging.

(Advances in chemistry series, 135)
"A symposium sponsored by the Division of Agricultural
and Food Chemistry at the 166th Meeting of the American
Chemical Society, Chicago, Ill., Aug. 30, 1973."

Includes bibliographical references.

1. Food—Packaging—Congresses.
I. Swalm, Charles M., 1918-   ed. II. American
Chemical Society. III. American Chemical Society. Di-
vision of Agricultural and Food Chemistry. IV. Series.

QD1.A355 no. 135 [TP374] 540'.8s [664'.09] 74-17150
ISBN 0-8412-0205-2   ADCSAJ 135 1-109 (1974)

Second Printing 1977

Advances in Chemistry Series

Robert F. Gould, *Editor*

FOREWORD

ADVANCES IN CHEMISTRY SERIES was founded in 1949 by the American Chemical Society as an outlet for symposia and collections of data in special areas of topical interest that could not be accommodated in the Society's journals. It provides a medium for symposia that would otherwise be fragmented, their papers distributed among several journals or not published at all. Papers are refereed critically according to ACS editorial standards and receive the careful attention and processing characteristic of ACS publications. Papers published in ADVANCES IN CHEMISTRY SERIES are original contributions not published elsewhere in whole or major part and include reports of research as well as reviews since symposia may embrace both types of presentation.

CONTENTS

Preface ix

1. Trends in the Design of Food Containers 1
   R. E. Beese and R. J. Ludwigsen

2. Glass Containers as Protective Packaging for Foods 15
   J. M. Sharf

3. Tinplate Containers for Packaging Irradiation-Sterilized Foods 22
   J. J. Killoran, E. Wierbicki, G. P. Pratt, K. R. Rentmeester,
   E. W. Hitchler, and W. A. Fourier

4. Compatibility of Aluminum for Food Packaging 35
   M. A. Jimenez and E. H. Kane

5. Packaging Food Products in Plastic Containers 49
   Charles A. Speas

6. High Nitrile Copolymers for Food and Beverage Packaging 61
   Morris Salame and Edward J. Temple

7. Flexible Packaging and Food Product Compatibility 77
   John E. Snow

8. Irradiation of Multilayered Materials for Packaging Thermoprocessed
   Foods 87
   John J. Killoran

9. Future Needs in Food Packaging Materials 95
   Seymour G. Gilbert

Index 103

PREFACE

A s society develops and technology increases, researchers are modifying old, accepted foods and introducing new products into the modern diet. Their years of constant research have brought us from the colonial ways of "eat off the vine" to the present use of preprocessed, modified, and synthetic foods.

Food packaging has similarly undergone radical changes. As the place of production grows farther from the urban centers where most of the food is consumed, the demands on food containers are greatly increased. Society used to be content to deal with natural food packaging, but now food must be shipped over long distances and stored for undetermined time periods and under often uncertain conditions.

The technologies of glass, tin, and aluminum containers are being improved, and the field of polymer containers is rapidly expanding. Now a food container may be rigid or flexible and may be made up of many combinations of films, layers, and coatings. It must be compatible with the food contained, must protect the product during processing, shipment, and storage, and must also satisfy marketing requirements for consumer acceptance.

Camden, N. J.   CHARLES M. SWALM
July 1974

PREFACE

Trends in the Design of Food Containers

R. E. BEESE and R. J. LUDWIGSEN

Material Sciences, Research, and Development, American Can Co.,
Barrington, Ill.

*Designs of metal cans for foods are continuously modified
to reduce cost and to improve container integrity and quality. Advances in container materials include steel-making
processes for tin mill products, corrosion resistant ETP (electrolytic tinplate) for mildly acid foods, tin free steel tin
mill products, and new organic coatings. Recent trends
in container construction result from antipollution legislation, new can-making technology, and public safety considerations. Water base and UV cured organic coatings
reduce pollution. Drawing, draw and ironing, cementing,
and welding provide alternate methods for making cans and
potentially upgrade container performance. Full inside
solder fillets in soldered sanitary cans also improve container
integrity and thus contribute to public safety. The soldered
sanitary can remains an important factor in preserving foods.*

Although soldered tinplate cans now dominate the processed food industry (*1*), changes are being made. The canning industry and
container manufacturers are responding to increased social and economic
pressures to change the traditional methods. Some of the most important
of these pressures are the efforts to protect the environment and the
concern over the public health aspects of canned foods. Renewed attention is being given to improved container integrity and safe canning
practices. New can-making materials and manufacturing techniques are
contributing to the solution of these problems. Recent changes in container construction permit the use of lower gage steel, lower tin coating
weights with improved corrosion resistance, beading of can bodies, and
increased use of inside organic coatings, which have all helped to minimize the cost of the tin can without reducing container quality or
integrity.

The technical aspects influencing these changes are reviewed in this paper. Discussion of these trends is limited to the steel-based materials. The current demand for easy-open ends for food containers has led to the development of many scored easy-open ends. This is a subject in itself and is not included in this discussion.

Materials

To appreciate the potential changes in food cans, it is necessary to describe briefly the steel-based materials used in modern can manufacturing operations. The tin can is made from a special grade of thin gage, low carbon, cold-rolled steel, which is generally referred to as a tin mill product. The base steel is coated with either tin, a chromium–chromium oxide system, or it is just cleaned and oiled. It may also be coated with organic coatings.

Electrolytic Tinplate. Much of the tin mill product is made into electrolytic tinplate (ETP). A schematic of an ETP cross section is given in Figure 1. The steel strip is cleaned electrolytically in an alkaline bath to remove rolling lubricants and dirt, pickled in dilute mineral acid, usually with electric current applied to remove oxides, and plated with tin. It is then passed through a melting tower to melt and reflow the tin coating to form the shiny tin surface and the tin-iron alloy layer, chemically treated to stabilize the surface to prevent growth of tin oxide, and lubricated with a thin layer of synthetic oil.

The tin coating on ETP can be purchased in 10 thickness ranges. Differentially coated plate, such as #100/25 ETP, is coated with 60×10^{-6} inch of tin on the #100 side and 15×10^{-6} inch of tin on the other side. The use of differentially coated ETP has markedly reduced the requirement for tin metal (2).

In plain tinplate cans for acid foods, tin provides cathodic protection to steel (3, 4). The slow dissolution of tin prevents steel corrosion. Many investigators (5–11) have defined this mechanism in detail and have shown that the tin dissolution rate is a function of the cathodic activity of the base steel, the steel area exposed through the tin and the tin–iron alloy layers, and the stannous ion concentration. Kamm et al. showed that control of the growth of the tin–iron alloy layer provides a nearly continuous tin–iron alloy layer and improves the corrosion resistance of heavily coated (over 45×10^{-6} in. tin) ETP for mildly acid food products in which tin provides cathodic protection to steel (12). The controlled tin–iron alloy layer reduces the area of steel exposed to the product. ETP with the controlled alloy is designated type K, and since 1964, #75 type K ETP has been used to provide the same protection as #100 ETP provided previously (13).

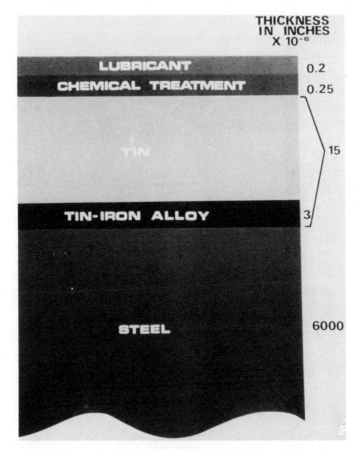

Figure 1.   Schematic cross section of 55 2CR tinplate (#25 tin coating)

Tinplate can be purchased in a wide range of tempers and thicknesses. Currently 17 different basis weights are available commercially, from 55#/BB (pounds per base box or 217.78 ft²) to 135#/BB. These weights range in nominal thickness from 6.1 to 14.9 mils.

The final thickness of the steel for tinplate is achieved by two processes. For conventionally reduced or single reduced plate, steel is annealed after cold reduction to restore ductility. The annealed coil is then temper rolled with only 1–2% reduction to make the final adjustment to its tensile strength, hardness, and surface finish. To reduce the cost of the lighter basis weight plates, double reduced, or 2CR, plate was introduced (*14*). A second 30–40% cold reduction is given steel after the anneal, which imparts a significant amount of cold work. This provides 2CR plate with generally greater hardness and tensile strength, a loss in ductility, and an increase in directionality. At first, these factors made 2CR plate more difficult to fabricate into containers. However, modification in manufacturing procedures has enabled this lower cost base steel to be used.

The chemistry of the base steel is carefully regulated to control both the physical properties and corrosion resistance (15). Recent changes in steel manufacture have generally benefited tinplate performance. Basic oxygen processes, which permit steel to be made at a faster rate, tend to produce low carbon steel with lower levels of residual elements. In general this is believed to improve corrosion resistance. However, in one case, there is a reduction in the corrosion resistance of steel for cola type carbonated beverages when the residual sulfur concentration is lowered from 0.035 to 0.018% (16, 17). In lemon–lime beverages, however, the lower sulfur levels improve corrosion performance. It is the copper/sulfur ratio which determines the corrosion resistance of steel for carbonated beverages.

The new continuous casting processes, in contrast to ingot cast products, provide tin mill products which are exceptionally clean and formable. The deoxidizing processes required for continuous casting involve either aluminum or silicon killing, which adds aluminum or silicon to the steel. Experience with type D steels indicates that the added aluminum will not cause a corrosion problem. Laubscher and Weyandt (18) have shown that the silicon found in silicon killed, continuous cast, heavily coated ETP will not adversely affect the corrosion performance of plain cans packed with mildly acid food products in which tin usually protects steel. The data on enameled cans is not definitive. Additional published data are required to determine whether or not silicon actually reduces the performance of enameled cans made from enameled, heavily coated, silicon killed, continuous cast ETP.

Tin Free Steel—Electrolytic Chromium-Coated. A less expensive substitute for tinplate, electrolytic chromium coated-steel, has been developed and is designated TFS–CT (tin free steel–chromium type) or TFS–CCO (tin free steel–chromium-chromium oxide) (19). This material can be used for many products where the cathodic protection usually supplied by tin is not needed. A schematic cross section is shown in Figure 2. Electrolytic, chromium-coated steel is made by electrolytically depositing a thin layer of metallic chromium on the basic tin mill steel, which is in turn covered by a thin passive coherent layer of chromium oxide.

Organic coatings adhere to the electrolytic chromium-coated steel surface exceptionally well. The surface is stable and does not discolor during baking of enamels. It is resistant to staining from products containing high levels of sulfide, such as meat, fish, and some vegetables.

It resists pinpoint rust formation before enameling and filiform corrosion after enameling.

TFS–CT or TFS–CCO is a primary material for cemented and welded beer and carbonated beverage containers (*20–22*) and can be used in sanitary food cans. It is currently used for ends on soldered sanitary food cans and is a candidate for drawn containers which do not require soldering.

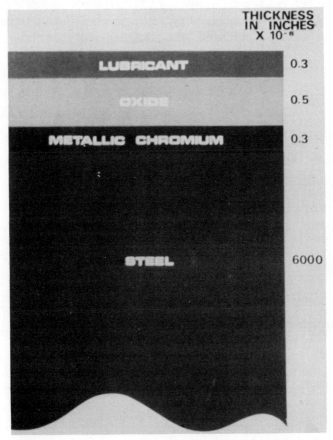

Figure 2.   Schematic cross section of tin free steel (TFS-CT)

Tin Free Steel—Can-Maker's Quality. CMQ (can-maker's quality) steel is the basic tin mill product. CMQ can be either single or double reduced steel. Rolling oils are removed, and the surface may or may not be passivated. A schematic cross section of passivated CMQ is shown in Figure 3. QAR (quality as rolled) 2CR plate is the basic 2CR tin mill product with the rolling oils on the surface. No further treatment is given. Figure 4 is a schematic cross section of QAR plate.

CMQ is commonly referred to as black plate and has borderline corrosion resistance both before and after enameling. Handling and storage before enameling must be carefully controlled to minimize pinpoint rust. Special organic coatings are required to control both internal corrosion and external filiform corrosion. They are colored to cover the brown appearance which forms during enamel baking. Undercutting corrosion resistance, as shown in Figure 5, is very poor. The chromate–phosphate TFS is a heavily passivated CMQ which is currently not being considered because of undercutting resistance and cost. Containers for dry products can be fabricated from CMQ.

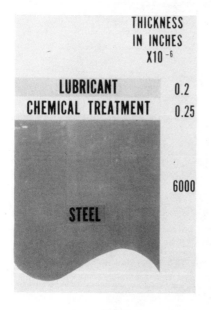

Figure 3.   Schematic cross section of TFS-CMQ plate

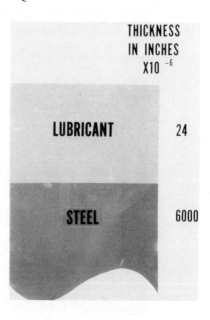

Figure 4.   Schematic cross section of TFS-QAR plate

QAR is being evaluated for cemented or welded beverage cans. Special organic coatings and manufacturing techniques are required because of the high level of residual rolling oils. Reasonable success has been achieved in making beer and beverage cans from QAR plate.

Organic Coatings. Organic coatings or lacquers protect the steel or tin from external or internal corrosion. The can interiors are coated to prevent undesirable reactions between the interior metal surface and the product. These reactions involve: (1) corrosion of the tin coating caused by oxidants in the product, (2) color or flavor loss by the product because of metal ion pickup, or (3) staining of the metal by sulfur-containing products. External inks and coatings are used to decorate the can, to minimize rusting, and to improve mobility.

Hundreds of coatings are available to the can maker. Coatings are selected for each use on the basis of which will give adequate performance at lowest cost (23). Recent advances in coating technology result

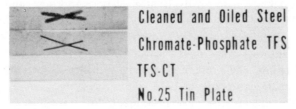

Figure 5. Undercutting corrosion resistance of enameled plate

from current application and usage requirements. For example, the new vended can for formulated food products requires enamel performance levels which could not be met by the oleoresinous coatings commonly used. The organosol and white sanitary coatings were developed to meet this need. The pigmented white coatings not only provide a pleasing aesthetic effect but also conceal underfilm staining produced by some products. In addition to organosol white coatings, white coatings based on epoxy–ester, acrylic, and polyester resins have been developed which meet sanitary food can requirements. Aluminum pigmented epoxy–phenolics and organosols have also been developed to conceal underfilm staining.

The mechanism which permits can coatings to prevent metal corrosion or staining has not been elucidated completely. Container coatings are only 0.2 or 0.3 mils thick, which is 1/10 to 1/100 as thick as conventional coatings used to protect tanks, pipes, or siding from atmospheric corrosion. Some investigators believe organic coatings combat corrosion by physical barrier, chemical inhibition, and/or electrical effects (24). Can coatings permit water absorption and diffusion, transport of ions such as hydrogen and chloride, and gas diffusion (25). These diffusion mechanisms suggest that corrosion can occur through continuous container coatings. Thus it is reasonable to conclude that when coating failures occur, it is because of one or all of these mechanisms.

Trends

There are several areas to consider when discussing the future of the food can. Anti-pollution legislation, new can-making technologies, and public safety aspects will have a pronounced effect on food container design.

Anti-Pollution Legislation. Anti-pollution legislation covers a broad area of social responsibility. For the can manufacturer it ranges from the requirements of the Clean Air Act of 1970 to reduce or eliminate contaminants which pollute the air to the ban-the-can type legislation as enacted by the state of Oregon. The latter type of legislation is directed mainly at deterring roadside litter of beer and beverage containers. The Clean Air Act has a significant effect upon the operations of industrial coating users. The requirements to reduce or eliminate organic emissions and noxious fumes from organic coating operations is of particular concern. As a result, the container coating industry is actively trying to develop alternatives to solvent base coating systems.

The following are the most prominent developments:

1. Water-base coatings for spray and roller coat application
2. Electrodeposition of water-base coatings
3. Heat cured high solids coatings
4. Radiation cured high solids coatings
5. Electrostatic sprayed powder coatings
6. Hot melt spray coatings

These technical developments and their merits have recently been well documented by R. M. Brick (26). The potential effect of these developments with regard to the ordinary hot-filled or steam-sterilized sanitary food container will take some time to discover because the use of organic coating materials on the inside of food containers is controlled by the Food and Drug Administration. Their guidelines restrict the compositional structure of coating materials and limit the amount of organic material which can be extracted from the coating by a food product (27). In general, container coatings for steam-sterilized food products must withstand the most stringent tests as described in these regulations.

Since most of the alternative coating systems above utilize polymer materials or adjuvants which are not acceptable food contact materials, a change in coating systems for the inside of food cans will probably be slow because of the testing required. Thus, the can maker will initially use incineration or adsorption of solvent emissions to comply with the pollution regulations. However, high solids, heat-cured sanitary, and C-enamels are now being evaluated to eliminate the need for pollution control. They have essentially the same chemical composition and properties as current coating materials but do not require control of the low amount of organic solvents released to the atmosphere during oven baking.

At present, the major effort to develop anti-pollution coating systems is devoted to outside lithographic decorating materials for beverage and non-food containers. Water-based, clear varnishes and white coatings are increasingly available for specific end uses. Significant advances are being made in using ultraviolet radiation as a nonpolluting curing method for container decoration. The inks and coatings used in the UV curing process are essentially 100% solids materials which are generally composed of acrylic or linear unsaturated polyester monomers combined with a suitable photosensitizer.

New Can-Making Technologies. The second area affecting container trends is the new can-making technology. Several manufacturing techniques are being considered which would compete with the conventional soldered tinplate sanitary container. These include drawing, drawn and ironing, and cementing and welding. The commercial success of the deep drawn aluminum food can is well known. Drawn and drawn and ironed processes for steel-based materials are being evaluated. The technology for replacing the soldered side seam with either cementing or welding techniques has been developed.

Advances in materials and new container construction techniques are usually evaluated with one- or two-year test packs. The time required to prove performance of new materials or container constructions slows development programs. However, several laboratory techniques are available which provide a reasonable estimate of container performance.

Figure 6.   Drawn cans for food

While each container manufacturer has developed proprietary tests, most are based on electrochemical techniques. Corrosion in enameled ETP or TFS cans can be evaluated using one of the available procedures (28, 29, 30). Corrosion performance of plain tinplate cans can be estimated using the Progressive ATC Test developed by Kamm (6, 7). These tests should speed the development of new containers.

DRAWN CANS. Drawn cans, shown in the Figure 6, are punched from a flat sheet of coated stock which has been lubricated with a solid lubricant such as wax (*31*). Drawing is probably the simplest method of container manufacture, requiring the fewest number of operations from sheet or coil stock to finished can. The presses used usually have multiple dies, permitting the manufacture of two or more cans with each press stroke. For shallow drawing this might be done in a single operation. For deeper drawing multi-stage draws are required. The final operation is performed on a beader, which beads and necks-in the bottom end to form the stacking feature. The process is suitable for TFS, ETP, and aluminum. The absence of a side seam and bottom end seam eliminates the problems associated with these areas and adds aesthetic appeal. However, the process places severe requirements on the metal surface and organic coatings. Some enamel coatings will meet the demands of this process, but care must be taken to preserve the integrity of the enamel coating which ultimately determines container performance.

Figure 7.   Drawn and ironed cans for food

DRAWN AND IRONED STEEL CANS. Drawing and ironing grew out of the drawing process. Figure 7 depicts D & I cans for foods. The drawn and ironed tinplate can is usually produced by drawing a cup from a coil of lubricated matte tinplate.

The need for matte tinplate is an important factor in the commercial success of the D & I steel-based beverage container (*32*). Drawn cups are fed into an ironing press which thins or irons the side walls, thereby increasing the height of the can. The side wall of the current tinplate D & I can for beer is reduced from 0.013 to 0.0045 inch by the ironing process. After ironing the can is trimmed, and lubricants are removed by washing. Washed, dry cans are decorated on the outside, baked, and then spray coated on the inside and baked. The finished can is necked and flanged. Beading strengthens the thin ironed side walls. This process is commercially applicable to both tinplate and aluminum and is presently used for beer and carbonated beverage cans (*33*).

There are several disadvantages to drawn and ironed cans for foods. The D & I can is used only with products suitable for enameled cans since there is not enough tin on the interior surface of a plain D & I can to provide cathodic protection to the large area of steel exposed through the ironed tin surface.

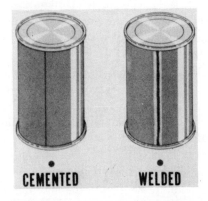

CEMENTED   WELDED

Figure 8.   Cemented and welded beverage cans

CEMENTED AND WELDED CANS. Beer and carbonated beverage cans, made by the now familiar cementing (*22*) and welding (*20*) processes, are shown in Figure 8. These processes could also be used for sanitary processed food cans. Enameled TFS materials are used for these cans. Corrosion performance of the enameled, cemented, and welded cans is similar to that of enameled soldered cans for products which do not require the cathodic protection usually supplied by the tin coating.

The cemented lap seam used in these cans is a sandwich of plate, organic coatings, and cement. The body is formed on a modified bodymaker, and the process is based on the control of heat input and removal. Welded cans are also made on a modified bodymaker. Coated TFS body blanks with bare edges are fed into the bodymaker where the margins are cleaned so that a uniform electrical resistance will be presented to the electric current provided to weld the side seam.

The service life performance of the cemented or welded enameled TFS container for many food products should be similar to the performance provided by an enameled #25 ETP soldered container. There is no increase in the cold work of the base steel nor is there any damage to the enamel coatings by the container forming processes. Low tin solder is quite inert, and thus the removal of the solder from the container by either the cementing or welding operations should not affect the container performance.

Public Health Aspects. The third factor affecting container trends involves public health. It is no coincidence that the new container-making processes which have been described also have a strong potential for upgrading can integrity for processed food products. The two-piece can (drawn or drawn and ironed) has only one end to be double seamed and has no soldered side seam. Since the open end of the can body is a smooth, continuous flange, there is less chance for false seams and recontamination during processing. Since there is no side seam, there is no solder, and hence, no chance for lead migration into the food. Welded or cemented can bodies also provide the same potential for eliminating concern over lead migration but without the advantage of end double seaming noted for two-piece cans.

Improved can integrity can also be provided for soldered tinplate cans. The recent trend is to flow solder all the way through the third fold of the side seam to provide a full fillet of low tin solder on the inside of the can (21). This ISF (inside solder fillet) sanitary can has a higher level of can integrity. The solder fillet gives additional creep and blowup strength and is more easily coated with organic side seam stripes. The electrochemical potential of low tin solder is such that any tin or steel exposed to the product will give cathodic protection to the low tin (98% lead) solder exposed at the side seam (34). The enameled sanitary can may also be soldered with pure tin solder. This special construction is known as the high tin fillet (HTF) can. Tin is available to provide cathodic protection to steel exposed through the breaks in the enamel coating because the cans are soldered with pure tin solder so that a 1/32 in. wide fillet of tin is exposed along the side seam (35, 36). This can is particularly suited for asparagus and some tomato products.

Conclusion

In spite of all these new container innovations, there are many situations where tinplate is required because of its corrosion resistance and ability to maintain product quality. Thus, it is believed that the soldered tinplate sanitary can will remain an important factor in preserving foods for many years to come.

Literature Cited

1. Coonen, N. H., Mason, S. I., "Recent Advances in Rigid Metal Containers," *Proc. Int. Congr. Food Sci. Technol.*, **3rd,** SOS/70, Washington, D.C. Aug. 9-14, 1970, pp. 589-594.
2. McKirahan, R. D., Connell, J. C., Hotchner, S. J., *Food Technol.* (1959) **13,** 228-232.
3. Kohman, E. F., Sanborn, N. H., *Ind. Eng. Chem.* (1928) **20,** 76-79.
4. Lueck, R. H., Blair, H. T., *Trans. Amer. Electrochem. Soc.* (1928) **54,** 257-292.
5. Kamm, G. G., Willey, A. R., *Corrosion* (1961) **17,** 77t-84t.
6. Kamm, G. G., Willey, A. R., "The Electrochemistry of Tinplate Corrosion and Techniques for Evaluating Resistance to Corrosion by Acid Foods," *Proc. Int. Congr. Metal. Corrosion*, 1st, London, April, 1961, pp. 493-503.
7. Kamm, G. G., Willey, A. R., Beese, R. E., Krickl, J. L., *Corrosion* (1961) **17,** 84t-92t.
8. Koehler, E. L., *J. Electrochem. Soc.* (1956) **103,** 486-491.
9. Koehler, E. L., Canonico, C. M., *Corrosion* (1957) **13,** 227t-237t.
10. Vaurio, V. W., *Corrosion* (1950) **6,** 260-267.
11. Willey, A. R., Krickl, J. L., Hartwell, R. R., *Corrosion* (1956) **12,** 433t-440t.
12. Kamm, G. G., Willey, A. R., Beese, R. E., *Mater. Prot.* (Dec. 1964) **3** (12), 70-73.
13. Hotchner, S. J., Poole, C. J., "Recent Results on the Problems of Can Corrosion," *Proc. Int. Congr. Canned Foods,* **5th,** Vienna, October 3-6, 1967, pp. 151-171.
14. Brighton, K. W., Riester, D. W., Braun, O. G., *Nat. Canners Ass. Information Letter No.* **1909,** 61-64 (January 31, 1963).
15. *Book ASTM Stand.* (1973) Part 3, A 623-672.
16. Beese, R. E., Krickl, J. L., Bemis, L. E., *Proc. Soc. Soft Drink Technol.* (1968), 57-80.
17. Mittelman, M. D., Collins, J. R., Lawson, J. A., "Corrosion Resistant Tinplate for Carbonated Beverage Containers," *Tech. Meet. Amer. Iron Steel Inst.*, San Francisco, Calif., Nov. 18, 1965.
18. Laubscher, A. N., Weyandt, G. N., *J. Food Sci.* (1970) **35,** 823-827.
19. Kamm, G. G., Willey, A. R., Linde, N. J., *J. Electrochem. Soc.* (1969) **116,** 1299-1305.
20. Chiappe, W. T., *Mod. Packag.* (Mar., 1970) **43,** (3), 82-84.
21. Eike, E. F., Coonen, N. H., "The Changing Food Container," *Int. Congr. Canned Foods,* **6th,** Paris, November 14-17, 1972.
22. Kidder, D. R., Kamm, G. G., Kopetz, A. A., *Amer. Soc. Brew. Chem., Proc.* (1967) 138-144.
23. McKirahan, R. D., Ludwigsen, R. J., *Mater. Prot.* (Dec. 1968) **7,** (12), 29-32.
24. Hartley, R. A., *J. Mater.* (1972) **7,** 361-379.
25. Koehler, E. L., "Corrosion Under Organic Coatings," *U. R. Evans Intern. Conf. Localized Corrosion*, Williamsburg, Va., Dec. 6-11, 1971, Paper **51.**
26. Brick, R. M., "New Can Coating Systems to Meet 1975 Air Quality Regulations," Protective Coatings Division of the Chemical Institute of Canada, Toronto and Montreal, Canada, March 21-22, 1973.
27. Ives, M., *J. Amer. Diet. Ass.* (1957) **33,** (4), 347-351.
28. Bird, D. W., Jones, B. R., Warner, L. M., *Bull. Inst. Nat. Amelior.* (1972) **23,** 128-153.
29. Reznik, D., Mannheim, H. C., *Mod. Packag.* (Aug. 1966) **39,** (12), 127-130.
30. Walpole, J. F., *Bull. Inst. Nat. Amelior.* (1972) **23,** 22-29.

31. Johns, D. H., "Manufacturing Methods for Metal Cans," Proceedings of a
    Seminar on Metal Cans for Food Packaging—Materials, Methods, Selec-
    tion Criteria and New Developments, Univ. of California, Los Angeles
    and San Francisco, Calif., November 6-7, 1968.
32. Bolt, R. R., Wobbe, D. E., U.S. Patent 3,360,157 (1967).
33. Kaercher, R. W., Mod. Packag. (Oct., 1972) 45, (10), 66-70.
34. Ass. Food Drug Offic. U.S., Quart. Bull. (1960) 24, 193-195.
35. Hotchner, S. J., Kamm, G. G., Food Technol. (1967) 21, 901-906.
36. Kamm, G. G., U.S. Patent 3,268,344 (1966).

RECEIVED October 1, 1973.

Glass Containers as Protective Packaging for Foods

J. M. SHARF

Glass Container Manufacturers Institute, Inc., Washington, D.C. 20006

The U.S. Food and Drug Administration regulates food additives derived from packaging materials, especially heavy metals, monomers, plasticizers, stabilizers, antioxidants, colorants, or other components. Soda-lime-silica container glasses are inert and qualified by the U.S. Pharmacopoeia as having low aqueous extraction and light transmission. This glass is an impermeable barrier to liquid, vapor, or gas transfer. Aqueous foods are generally acidic, extracting minute amounts of the alkaline oxides, soda, and calcia from a container. Simultaneously, an adherent hydrated silica film is formed which limits the depth of reaction. Representative aqueous data show minor elements present only in ppb. On the basis of inertness and barrier characteristics, glass containers are found superior for hermetic packaging of foods.

The purpose of food packaging is clearly stated by the Food Protection Committee (FPC) of the National Academy of Sciences (1). The food package is to protect the contents during storage—both before sale and in the home—from contamination by dirt and other foreign material; infestation by insects, rodents, and microorganisms; and loss or gain of moisture, odors, or flavors. Frequently deterioration is controlled by preventing contact with air, contaminating gases, or light. Because the packages are closely associated with food, they must contribute little, if any, acceptable, harmless, incidental additives which originate in the packaging and are transferred to the food mechanically or by solution, extraction, or decomposition. These often unanticipated additives have been long recognized and are closely regulated by the U.S. Food and

Drug Administration (FDA) which is responsible for the safety of the food supply.

Early discussions with the FDA leading to the present regulations were concerned with the specific applications of the basic Food and Drug Law of 1938. Interpretations were derived from long experience with inherently harmless containers of glass, wood, and some metals. In the FDA Code of Federal Regulations (CFR), a specific section deals with food additives from containers (2). Test procedures are given which specify extractant, time, and temperature of exposure. These routines have little or no extraction effect on glass containers because of their inertness. However, there appears to be growing restriction in the interpretation of this section.

The underlying cause for the restriction in the CFR has been the increasing diversity and complexity of packaging materials, many of which contain substances quite foreign to food products. Some of these substances have been questioned as to long term physiological effect on the consumer since they are often incidental, unanticipated additives to the contained foods during storage. Questions arise as to the Pb in solders used to join side-seams; metallic organic complexes of Sn, Zn, Cd favored as polymer stabilizers; as well as a variety of organic molecules intended as plasticizers, antioxidents, colorants, and related agents. The plastics themselves may release unreacted monomers. This is caused by differing solvent actions of the various foods and beverages.

The continuing concern in the U.S. and Europe with incidental additives from packaging has been clearly stated by Golberg who has participated extensively in the discussions of the FPC (3). He also reviewed the so-called Frawley proposal which presented the concept that any substance used as a functional component in food packaging (other than pesticides and heavy metals) at a level of 0.2% or less could not attain an unsafe level in the food. However, Golberg has found that this concept is not acceptable and that in the prevailing circumstances, few conclusions will be reached regarding the acceptability of these incidental additives by the authorities. Adding to this complication is the problem of determining and measuring the quantities of the compounds migrating into foods from any single packaging material.

The inertness of contemporary soda-lime-silica glass is so great that investigators rarely give it second thought. Therefore, it is important to understand in detail the composition of the glass as well as its low order of extraction by aqueous foods. As a rigid container, glass does not require plasticizers. It is composed exclusively of stable earthen oxides and does not employ stabilizers or antioxidants. Heavy metals are not a component, and since it is formed in one continuous container structure, solders containing lead are not used.

Most packaged foods or beverages (as well as drugs and cosmetics) are neutral or acidic. Therefore, the extraction of the component earthern oxides from glass containers by neutral and acidic aqueous solutions is considered below. Since glass is quite abrasion resistant and is impervious to fats or oils, the action of dry or fatty foods is not discussed. The transparency of glass is an advantage in identifying the contents, but if restriction of active wavelengths in the near ultraviolet though the visible range is desirable, amber glass may be utilized. The rigid glass container is also used to retain vacuum or positive pressure with hermetic functioning closures.

Composition of Soda-Lime-Silica Container Glass

The major ingredient of this glass is selected sand, SiO_2, which is fluxed and melted in large tonnage at commercially attainable temperatures by the addition of soda ash, Na_2CO_3, which upon firing becomes Na_2O. This two-component glass is clear but has no resistance to the hydrating action of water, and the conventional water-glass solution would easily form. The stability against water results from adding limestone, frequently dolomitic, which in the melt furnishes the divalent elements as oxides CaO and MgO. The resistance and mechanical forming properties are improved by a lesser amount of Al_2O_3. The resulting glass is not a true chemical compound, but more a lattice or micelle of oxygen atoms in which there is a statistically random distribution of the positive ions. In these compositions, the O_2 content of glass approaches 50% by weight.

The theoretical mechanism of the action of water on such glass has been fully considered by Douglas and El-Shamy (4). The most aggressive solution is double-distilled water at neutrality. The effect of dilute acidic solutions is much less, the main action being the extraction of alkali (Na^+) ions which are replaced by hydrogen ions. The result is a surface zone where the glass is depleted of sodium. Although traces of extracted silicates may appear in the solution, this resident dealkalized layer becomes a barrier to further ionic diffusion, reducing the extraction to a very low terminal rate. The aqueous phase of the majority of food products is acidic.

The U.S. Pharmacopoeia (USP) Standards

The USP stipulates a test and limits for alkali extraction from container soda-lime glass (5). Specially prepared double-distilled water is used to extract the glass for one hour at 121°C in a steam autoclave on a strict cycle program. An aliquot of the extract is then back titrated with

$50N$ H_2SO_4 using methyl red indicator. Soda-lime-silica glass qualifies as Type III with a limit of 8.5 ml of the back titrating acid per 50 ml of extract. The current routine employs powdered glass ($+50$, -20 sieve) because it is considered more representative of the character of the glass than the inner surface of a particular bottle. However, in the past the test has applied to bottles, and the pass level was less than 30 mg/l. total extractables expressed as sulfates for the typical small bottle. Further, the exposure to accelerated conditions of 121°C for one hour in an autoclave is considered to represent more than one year of storage at room temperature and probably almost a typical two-year period.

The accelerated test for soda-lime glass containers using the steam autoclave has been reviewed fully by Bacon and Burch, whose findings form the basis for current procedures (6, 7). The relationship of the various compositions of all silicate glasses has been compared in the more recent study by Bacon (8, 9, 10). Emphasis is placed on the depressing effect of various ions in water which extensively slow the reaction. Distilled water is considered the most aggressive except for alkaline solutions which tend to remove the diffusion controlling, adherent, hydrated silica layer producing minute flakes. This effect is occasionally seen in stored reagent bottles of ammonia or caustic.

Analysis of the Aqueous Extract

The characteristic aqueous extract from soda-lime glass has been identified in extensive analytical studies by Poole (11). These are shown in Table I. An empirical ratio of extract ppm divided by component percentage gives an approximation of the diffusion rates through the hydrated interfering silica layer. The occurrence of the layer is indicated by the proportionally low ratio for SiO_2 in the extract. Among the major constituents, Na appears the most active (except for the K_2O present in a small amount, probably derived from feldspar the source of Al_2O_3). This particular glass has a contemporary range for these first two major components, but the CaO and MgO level derived from dolomitic limestone has been selected to show the trend of decreasing ion mobility. The very low ratio for Al_2O_3 indicates low mobility and the apparent blocking action in the silica film. This explains the lower rate of extraction found in soda-lime glasses which contain alumina.

Iron is associated with silica sand, usually as a light surface stain on the grains. Amber glass develops ionic color centers or complexes of Fe-S-C added to the batch as iron sulfide and powdered anthracite. Although the Fe content be four or five times that shown in the example in Table I, it appears to be bound in the complex so that no greater extraction occurs with the S and C. Titanium is associated with sand as ilmenite or rutile, and the other elements appear to be present at background levels associated with the glass batch materials. Cullet is selected, broken soda-lime glass which originates in the plant or is purchased on the outside. It is used for approximately 20% of the composition and appears to have little effect on the proportion of extractables. There has been no observable "multiplier" effect on any of these residual elements when recycled cullet is used instead of earthen batch materials alone.

Table I.   Relationship of Flint Glass Composition and Aqueous Extract[a]

Glass Composition, %		Aqueous Extract[b], ppm	Ratio, ppm/%
SiO_2	71.6	20.0	0.28
Na_2O	13.5	7.0	0.52
CaO	10.3	5.0	0.48
MgO	2.50	1.0	0.40
Al_2O_3	1.25	0.07	0.06
BaO_2	0.35	0.003	0.08
K_2O	0.23	0.2	0.87
Fe_3O_4	0.04	0.02	0.50
Totals	99.77	33.293	

[a] The above accelerated extraction procedure employed special double distilled water exposed to the soda-lime glass surfaces for 2 hours in a steam autoclave at 121°C. This schedule is considered to represent 3–4 years of room temperature shelf life of a typical container filled with distilled water. If the container and water were a product sterilized at 121°C for an hour and then stored, this schedule would represent a shelf life of at least 2 years. Acidic food products would extract one-quarter to one-half this amount because of ion interference.

[b] Analysis by atomic absorption showed other residual elements present in minute amounts in the extract: TiO_2, 7.0 ppb; AsO_2O_3, 1.0 ppb; PbO, 0.3 ppb; and both Co_3O_4 and CdO less than 0.1 ppb.

External Surface Protective Coatings

Protective coatings are frequently used on glass containers to increase their performance and flow through high speed filling lines. They are applied only to the exterior surfaces either by cold vapor fogging or hot surface reaction at one or both ends of the annealing lehr. The cold surface coatings may be minute, invisible amounts of lubricious food grade stearates, oleates, or polyethylene. The hot end coatings develop by exposing containers which are still hot from forming machines to dry-air diluted vapors of tin chloride or titanium chloride. An oxide film of either metal forms on the exterior glass surface. The amount of film is controlled to prevent iridesence which can occur when the coating is 80 microns thick. These oxide surface coats are somewhat harder than the glass, form a tightly adherent bond with the cold end coatings, and mechanically protect as well as lubricate the contact surfaces. Being restricted to external surfaces, they are remote from the contents.

Closures

Closures for glass containers are commonly a one-piece metal shell with threads or lugs for attachment to the glass container opening (since homogeneous glass stoppers are of limited use). The exposed metal surfaces are coated with impermeable continuous films of fully cured varnishes or lacquer. The typical, resilient sealing gasket used for fully hermetic closures is a stabilized elastomer, presenting a minimum surface to the contained product. Although the surface coatings often are similar to those used in the interiors of metal containers, the exposed closure area is a small fraction of that for an all-metal container so that extractables, if any, are minor by comparison.

Light Transmission

Light transmission characteristics of soda-lime glass depend on the absence or development of ionic color centers involving a fractional percentage of iron and its complexing. The clear or flint glass is low in iron and other metallic elements as Table I indicates. Thus it is transparent which is advantageous in food containers. The typical transmission curve for flint glass shows virtually zero transmission of the near ultraviolet at wavelength of 290 $m\mu$, rapidly ascending to approximately 90% at about 40 $m\mu$, and continuing into the infrared. Some food and drug products may show specific, characteristic wavelength absorption bands in the 290–450 $m\mu$ range, and energy absorbed at these bands may initiate some changes in the product. Hence in recognition, the USP XVIII and earlier editions have stipulated for a "light resistant container" the maximum percentage of light transmission of a closure sealed glass container is "not to exceed 10% in any wavelength 290 to 450 $m\mu$." This qualification is readily attained by amber glass. A simple alternative is placing a paper label of adequate area on a flint glass container to act as a light barrier.

Summary

The contemporary soda-lime-silica glass is formed by fusing selected sand, soda ash, and limestone, with lesser alternate material such as feldspar. When such a glass is given an accelerated test by extraction with double distilled water for 2 hours in a steam autoclave at 121°C, a tightly adhered hydrated silica barrier forms on the glass interface which markedly reduces the diffusion of positive ions. In descending order of appearance in the aqueous extract were Na_2O, CaO, MgO, and Al_2O_3. The test is considered to be the equivalent of 3 to 4 years of shelf life for distilled water in the glass container. The total extract of other lesser oxides, was 33 ppm. An acidic solution would reduce this level to one-quarter to one-half this amount. Analysis by atomic absorption showed other residual elements present in minute amounts in the extract: TiO_2, 7.0 ppb; As_2O_3, 1.0 ppb; PbO, 0.3 ppb; and both Co_3O_4 and CdO less than 0.1 ppb. The glass can be made amber color, effectively reducing transmission in the 290-450 mμ range to less than 10%. As a packaging material glass has superior performance characteristics compared with metallic, plastic, composite, or other barrier structures particularly for processed foods requiring hermetic containment.

Literature Cited

1. Food Protection Committee, "The use of chemicals in food production, processing, storage, and distribution," Natl. Acad. Sci.-Natl. Res. Council, Washington, D.C., 1973.
2. U.S. Food & Drug Administration, Code of Federal Regulations. Title 21—Food and Drugs, Part 121—Food Additives Subpart F., par. 121.2500, "Food additives resulting from contact with containers or equipment, and food additives otherwise affecting food," 1972.
3. Golberg, L., "Trace chemical contaminants in food; potential for harm," *Fd. Cosmet. Toxicol.* (1971) **9**, 65.
4. Douglas, R. W., El-Shamy, T. M. M., "Reaction of glasses with aqueous solutions," *J. Am. Ceram. Soc.* (1967) **50**, 1.
5. U.S. Pharmacopoeia XVIII, "Light transmission; Chemical resistance-glass containers," Easton, Pa., p. 923, 1970.
6. Bacon, F. R., Burch, O. G., "Effect of time and temperature on accelerated chemical durability tests made on commercial glass bottles," *J. Am. Ceram. Soc.* (1940) **23**, 1.
7. *Ibid.* (1941) **24**, 29.
8. Bacon, F. R., "The chemical durability of silicate glass," *The Glass Ind.* (1968) **49**, No. 8, 438.
9. *Ibid.* No. 9, 494.
10. *Ibid.* No. 20, 554.
11. Poole, J. P., (1973) Brockway Glass Co., Brockway, Pa., private communication.

RECEIVED October 1, 1973.

3

Tinplate Containers for Packaging Irradiation-Sterilized Foods

J. J. KILLORAN and E. WIERBICKI

U.S. Army Natick Laboratories, Natick, Mass. 01760

G. P. PRATT, K. R. RENTMEESTER, E. W. HITCHLER, and W. A. FOURIER

American Can Co., Barrington, Ill. 60013

The reliability of the commercially available tinplate container was determined for packaging irradiation-processed foods. Eight enamels, three end-sealing compounds, two tinplates, and the side-seam solder were irradiated with 3.0–4.0 Mrad and 6–7.5 Mrad at 5, −30, and −90°C. The epoxy phenolic enamel was the preferred enamel. There were minimal extractives from this enamel in the presence of three food-simulating solvents. The preferred end-sealing compound was the blend of cured and uncured isobutylene–isoprene copolymer. Component testing of the tinplate and solder showed that the gamma radiation, even at −90°C, did not transform the beta tin, or silvery form, to the alpha tin, or powdery form. In a small-scale production test, the tinplate container was completely reliable for packaging irradiation-sterilized beef and ham.

At first, the program which investigated the packaging of irradiation-processed foods, concentrated on the most advanced type of container, the tinplate can. It had performed successfully for a century as a container for thermoprocessed foods. However, as a container for the irradiation-processed foods, its physical, chemical, and protective characteristics had to be evaluated, including the effects of radiation on enamels and end-sealing compounds. This container was satisfactory for packaging foods that were irradiation sterilized while unfrozen (*1, 2*).

With the advent of irradiation processing of frozen foods to maintain acceptable quality in beef, ham, pork, and chicken, questions were posed as to whether the metal can would perform satisfactorily. What is the effect of radiation and/or low temperature on the interior enamels and end-sealing compounds? Does the combination of radiation and the low temperature promote the conversion of tin from the beta, or silvery form, to the alpha, or powder form, rendering the tin coating ineffective in protecting the base steel of the tinplate?

This paper describes the work that was performed to answer these questions, *inter alia,* including

(1) A screening study for evaluating and selecting components of the tinplate container—tinplate, enamel, end-sealing compound, and side-seam solder—which were irradiated at designated doses and temperatures

(2) An extractive study of one can enamel in the presence of food-simulating solvents to determine how gamma radiation from a cobalt-60 source altered the nature and amount of extractives of this enamel.

(3) A performance test of cans of irradiation-sterilized meat products (3).

Experimental

Irradiation Conditions. The gamma (cobalt-60) radiation facility and the source calibration are described by Holm and Jarrett (4). Irradiation doses were 3–4 Mrad and 6–7.5 Mrad at 9×10^2 rads per second for the screening study. Irradiation temperatures were 5, −30, and −90°C. The gamma source was calibrated with the ferrous sulfate–cupric sulfate dosimeter.

Enamels. Eight commercial enamels, listed in Table I, were applied by roll coating to panels (25×76 cm) of 43 kg, Type MR-TU, No. 25 electrolytic tinplate. These panels were cut into strips (10×25 cm) and tested for flexibility before and after irradiation with the General Electric impact apparatus by the reverse impact method. The impactor

Table I. Enamels Coated on Tinplate[a]

Enamel	Dry Weight (mg/cm²)
1. Polybutadiene	390
2. Polybutadiene with zinc oxide pigment	495
3. Epoxy phenolic with aluminum pigment	416
4. Epoxy-wax and butadiene–styrene copolymer with aluminum pigment	442, 78
5. Epoxy-wax with aluminum pigment	442
6. Oleoresinous	598
7. Phenolic	156
8. Oleoresinous with zinc oxide and epoxy with aluminum pigment	598; 234

[a] 43 kg (95 lb), Type MR-TU, No. 25 (5.6 g/m²).

was dropped onto the strips from a height of 76 cm. Strips were graded for percent elongation in the range of 0.5–60%. Enamel adhesion to the tinplate was determined by scribing and taping the enameled test specimens with cellophane tape.

End-Sealing Compounds. The commercial end-sealing compounds used in this study were:

Compound A—A blend of cured and uncured isobutylene–isoprene copolymer

Compound B—A blend of polychloroprene and butadiene–styrene copolymer

Compound C—A blend of polychloroprene and uncured isobutylene–isoprene copolymer

Strips of tinplate (2.5 cm wide) coated on one side with the test enamels were dipped in solutions of the end-sealing compounds with a Fischer-Payne dip coater. The viscosities of the solution were adjusted so that the dried compound was 7.5×10^{-2} cm thick. Tests were performed for cohesion, adhesion, and brittleness of the end-sealing compounds coated on the various enamels. Cohesion is defined here as a measure of the combined elongation–elasticity property of the end-sealing compound. Adhesion is a measure of the adhesive strength between the end-sealing compound and the enamel. Both cohesion and adhesion were determined by manually picking at the end-sealing compound with a dissecting needle. A numerical grading scale between 1 and 10 was used to indicate the effect of irradiation and/or temperature. The effect of temperature (5, −30, and −90°C) on the elasticity of the unirradiated test specimens was determined by bending the specimens 90° over a 6.4 mm glass rod. An arbitrary numerical scale was used as the basis of analysis. Rigidity changes of the end-sealing compounds caused by the irradiation treatment were determined by torsional braid analysis. In this test a glass braid coated with an end-sealing compound is suspended vertically, and a weight is hung at the lower end to form a torsional pendulum. The period of rotational oscillation of the pendulum measures changes in the rigidity of the end-sealing compound coated on the braid (5).

Tinplate and Solder. The tests were carried out to determine the effect of low temperature irradiation on the metallurgical properties of the tinpalte, solder, and soldered lap joints. Two types of tinplate were used: 43 kg (95 lb), Type MR-TU and 43 kg (95 lb), Type MR-T2, both coated with No. 25 electrolytic tinplate. The test specimens were 20 × 20 cm panels.

The solder (2% tin–98% lead) was molded into test specimens with a 1.27 cm diameter reduced section. The soldered lap joint specimens were prepared from the tinplate and solder. A solder flux was applied to each joint, the clamped specimens were dipped into molten solder, and the joint was obtained through capillary action. The tensile specimens with a 1.27 cm wide reduced section were prepared from the soldered lap joints. Tensile tests were carried out with the Instron universal testing machine at a cross-head speed of 0.5 cm/minute. Impact tests were conducted on a Plas-Tech universal tester at a cross-head speed of 12.7×10^3 cm/minute. The corrosion resistance of the tin coatings was determined from the iron-solution values (ISV) and the alloy-tin-couple (ATC) method (6). The microstructure of the tinplate was analyzed with the scanning electron microscope (SEM) and the transmission electron microscope (TEM) (7). With SEM, the tin–iron alloy was examined directly on the plate surface after stripping the tin. TEM required examination of a carbon replica. The cast solder was examined with TEM.

Extractives from Enamel. The nature and concentration of extractives from the irradiated epoxy phenolic enamel (coated on tinfoil) were determined by chemical and microanalytical techniques and compared with the extractives from the same, but unirradiated, enamel in contact with the solvent under similar storage conditions (1). The food-simulating solvents were demineralized distilled water, 3% acetic acid, and n-heptane. Both the water and acetic acid extractives were treated with chloroform to produce a chloroform-soluble fraction containing most of the organic components and a chloroform-insoluble fraction containing most of the inorganic components. The aqueous solvents were stored after irradiation for six weeks at 38°C while the n-heptane was stored for four hours at 21°C. The irradiation dose was 4.7–7.1 Mrad at 21–40°C and at -30 ± 10°C. The irradiation-induced changes in the enamel and identification of extractives were determined by infrared spectroscopy using a Beckman IR 10 grating spectrophotometer.

Results and Discussion

Enamels. The flexibility grades for the eight enamels (Table I) that were irradiated with 3–4 Mrad and 6–7.5 Mrad at 5, −30, and −90°C are shown in Table II. These data indicate that the epoxy-based enamels showed the best initial flexibility at −90°C and maintained their flexibility after irradiation. The preferred enamels were the epoxy phenolic with aluminum pigment, epoxy-wax and butadiene–styrene copolymer with aluminum pigment, and epoxy-wax with aluminum pigment. Tinplate adhesion before and after irradiation was satisfactory for the eight enamels.

End-Sealing Compounds. The qualitative cohesion testing of the three end-sealing compounds with the dissecting needle showed that

Table II.   Flexibility of Irradiated Can Enamels

Enamel	Elongation at 5°C [a]			Elongation at −30°C [a]			Elongation at −90°C [a]		
	Control	4 Mrad	7.5 Mrad	Control	4 Mrad	7.5 Mrad	Control	4 Mrad	7.5 Mrad
1	40	40	40	20	40	40	10	5	5
2	20	20	20	20	20	20	5	5	5
3	60	60	60	60	60	60	20	10	10
4	60	40	60	60	60	60	20	10	10
5	20	40	40	60	60	60	20	10	10
6	40	40	40	20	20	20	5	5	5
7	20	20	20	20	20	20	10	5	5
8	60	40	60	20	20	20	5	5	5

[a] Expressed as percent elongation.

Compound A was affected most by the irradiation, Compound B least, and Compound C intermediate. The effect of the irradiation on cohesion increased with increasing irradiation dose and temperature. The isobutylene–isoprene copolymer in Compound A and Compound C degrades during irradiation (9), becoming softer after irradiation. Since the seam of a can is formed before irradiation, some softening of the compound in the seam is not detrimental to the integrity of the seam.

The three irradiated end-sealing compounds had good adhesion to all the enamels except Enamel 5, the epoxy-wax enamel with aluminum

Table III.   Effect of Irradiation on Cracking and Adhesion of End-Sealing Compounds[a]

Enamel	Compounds[b]					
	A		B		C	
	Cracking	Adhesion	Cracking	Adhesion	Cracking	Adhesion
1	3	1	7	6	5	7
2	4	2	7	7	6	6
3	4	3	8	6	4	4
4	5	4	6	8	4	5
5	4	4	7	7	5	6
6	4	3	7	7	6	6
7	5	4	7	7	6	7
8	5	4	7	8	6	5

[a] Irradiated at 6 to 7.5 Mrad at −90°C.
[b] Grading Scale:

Rating	Cracking	Adhesion
0	none	no adhesion loss
3	slight	slight adhesion loss
5	moderate	moderate adhesion loss
7	severe	severe adhesion loss
10	shattered	100% adhesion loss

pigment. When bent 90° over a 6.2 mm glass rod at −90°C, the order
of cracking resistance and adhesion to enamels for the unirradiated and
irradiated end-sealing compounds was: Compound A > Compound C >
Compound B, as summarized in Table III.

Table IV shows the data on rigidity changes of the end-sealing compounds at two dose levels. Rigidity was determined by torsional braid
analysis (5). These data indicate that the blend of cured and uncured
isobutylene–isoprene copolymer was softened most by the irradiation
treatment, the blend of polychloroprene and butadiene–styrene copolymer softened the least, and the blend of polychloroprene and the uncured
isobutylene–isoprene copolymer was intermediate. Increasing the irradiation dose from 3–4 Mrad to 6–7.5 Mrad decreased the rigidity of the
three end-sealing compounds. The irradiation temperature did not significantly influence rigidity.

Table IV.   Rigidity of End-Sealing Compounds After the Irradiation

Irradiation Conditions		Relative Rigidity[a]		
Dose (Mrad)	Temperature (°C)	A	B	C
3–4	5	0.81	0.85	0.78
6–7.5		0.52	0.78	0.64
3–4	−30	0.63	0.80	0.75
6–7.5		0.46	0.76	0.64
3–4	−90	0.71	0.92	0.77
6–7.5		0.65	0.76	0.71

[a] Relative rigidity = P_o^2/P^2 where P_o is period of pendulum of the control, and P is period of pendulum after irradiation. A value of 1 indicates no change; less than 1 indicates softening.

Table V.   Relative Rigidity of Unirradiated End-Sealing Compounds

	Relative Rigidity[a]		
Temperature (°C)	A	B	C
5	1.57	2.15	1.87
−30	5.12	26.6	11.4
−90	138.0	57.2	32.5

[a] Relative rigidity = P_o^2/P^2 where P_o is period of pendulum at 27°C, and P is period at test temperature.

The relative rigidities of unirradiated compounds at various temperatures with 27°C as standard are shown in Table V. The lower the
temperature, the higher the relative rigidity. The patterns of the rigidity
changes with decreasing temperature were different with each compound.
Compound A showed less change than B and C down to −30°C, but
more change at −90°C. Compound C showed more change than Compound A at −30°C, but less at −90°C. Compound B showed most

change at 5 and −30°C and ranked between A and C at −90°C. These relative rigidities show the rigidity change of the individual compound, since each braid is compared only with itself at different temperatures.

The rigidity changes caused by irradiation would not preclude the use of the three end-sealing compounds. Since the irradiation takes place after the seam is formed in a tinplate container, the end-sealing compound is distributed in the seam and a softening of the compound should not affect sealing performance. However, in selecting an end-sealing compound for a container for irradiation-sterilized foods, the overall data on adhesion, cohesion, rigidity indicate that the Compounds A and B would be preferred. Compound A had the best low temperature characteristics, and Compound B was the least affected by irradiation.

Tinplate and Solder. Metallurgical studies were performed to determine the effect of irradiation at low temperature on the corrosion resistance of tinplate and on the mechanical properties and microstructure of tinplate and side-seam solder of the tinplate container. The area of major interest was the effect of low-temperature irradiation on the possible conversion of the tin from the beta form to the alpha form. In the case of pure tin, the transition occurs at 18°C. It was feared that low-temperature irradiation would create dislocations in the crystal lattice of tin and enhance the conversion of tin from the silvery form to a powdery form rendering the tin coating ineffective in protecting the base steel. Tin used for industrial consumption contains trace amounts of soluble impurities of lead and antimony to retard this conversion for several years.

Table VI summarizes the results of tension tests on tinplate irradiated at 6–7.5 Mrad. Radiation had no apparent effect on the tensile

Table VI.   Effect of Radiation on Tensile Properties of Tinplate

Dose (Mrad)	Temperature (°C)	Yield Strength[a] (MPa)	Tensile at Failure[a] (MPa)	Elongation[b] (5%)	
				Tensile	Impact
(Control)	—	472	472	8.7	5.3
6 to 7.5	5	482	487	8.0	3.8
6 to 7.5	−30	473	473	9.3	7.2
6 to 7.5	−90	478	479	10.2	6.0

[a] Transverse to rolling direction.
[b] Tensile, 8.4×10^{-2} m/s; impact 2.1 m/s.

properties of the tinplate since the minor variations in tensile values reflect experimental error. Impact ductility data, in particular, was significant because it indicated that no embrittlement occurred in the tinplate as a result of the low-temperature irradiation at −30 and −90°C. Metallographic examination showed that the base steel was not affected by the low-temperature irradiation. Figure 1 shows the typical micro-

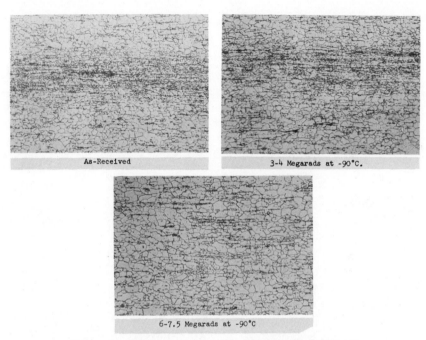

Figure 1.   Microstructure of the MR-TU steel (500 ×)

structures of spheroidal carbides and fine ferritic grain size before and after the irradiation for 43 kg (95 lb), Type MR-TU steelplate.

Table VII shows the data on the effect of the low-temperature irradiation on the tensile properties of cast 98–2 solder (98% lead–2% tin). These data indicate that the radiation had no effect on the tensile properties of the commercial solder which is used for the side seam of tinplate containers. Metallographic examination confirmed the absence of change in the microstructure of the solder after irradiation. (Figure 2).

The peel strength of soldered lap joint specimens was also not affected by irradiation. For example, the initial peel strength of a lap joint, fabricated from the 43 kg (95 lb) MR-TU, No. 25 tinplate, and the

Table VII.   Effect of Radiation on Tensile Properties of Solder

Dose (Mrad)	Temperature (°C)	Yield Strength (MPa)	Tensile at Failure[a] (MPa)	Elongation (%)
(Control)	—	9.5	18.4	54
6 to 7.5	5	8.4	17.9	41
6 to 7.5	−30	8.2	17.9	45
6 to 7.5	−90	8.3	18.3	53

[a] Tensile, 0.2% offset; elongation, in 5.1 cm.

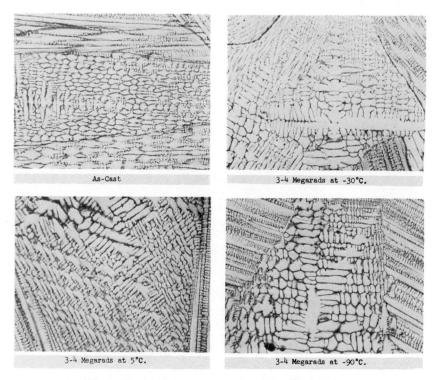

Figure 2.   Microstructure of the cast solder (150 ×)

98–2 solder, was 1.58×10^4 N/m and 1.55×10^4 N/m after irradiation at 6–7.5 Mrad at −90°C.

Changes in corrosion resistance of the electrolytic tin coatings were determined by the iron-solution-value test and the alloy-tin-couple test (6). Corrosion resistance data for the 43 kg (95 lb), Type MR-TU, No. 25 tinplate are presented in Table VIII. These data show that there is no significant difference in the corrosion resistance of the unirradiated

Table VIII.   Effect of Radiation on Corrosion Resistance of Tinplate

Radiation Conditions		*Alloy–Tin Couple*		*Iron Solution (µg Fe)*
Dose (Mrad)	*Temperature (°C)*	*Top*	*Bottom*	
(Control)	—	0.30	0.24	11
3 to 4	5	0.28	0.21	12
3 to 4	−30	0.24	0.19	10
3 to 4	−90	0.28	0.20	14
6 to 7.5	5	0.28	0.23	11
6 to 7.5	−30	0.24	0.20	14
6 to 7.5	−90	0.21	0.19	11

and irradiated tinplate. The scanning electron and transmission electron microscopic examination of the iron–tin alloy showed no evidence of change as a result of irradiation. The alloy shown in the photomicrographs of Figures 3 and 4 is typical of the continuous structure found on steelplate with superior corrosion resistance and indicates that the iron–tin alloy crystals provide remarkably complete coverage of the base steel.

As-Received

3-4 Megarads at -90°C.

6-7.5 Megarads at -90°C.

Figure 3. SEM photomicrographs of iron-tin alloy of detinned tinplate (20,000 ×)

The metallurgical experiments showed that the beta-alpha transition of the tin coating did not occur at irradiation doses of 3–5 Mrad and 6–7.5 Mrad at 5, −30, and −90°C and that the tensile properties, impact ductility, peel strength of soldered lap joints, and microstructure of commercial tinplate and solder were not affected by the irradiation conditions that are used in the sterilization of meat products.

Extractives from Can Enamels. Earlier work reported by Pratt (*1*) showed that in a comparison between irradiation processing and thermal processing, no significant differences were found in the amount of extractives obtained from three commercial can enamels—epoxy phenolic, polybutadiene, and oleoresinous—in the presence of three aqueous

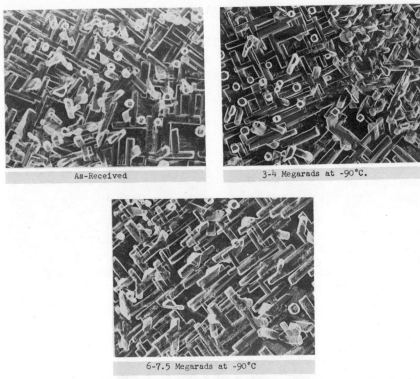

As-Received

3-4 Megarads at -90°C.

6-7.5 Megarads at -90°C

Figure 4. TEM photomicrographs of iron-tin alloy of detinned tinplate (20,000 ×)

solvents—simulating neutral, acid, and fatty foods. The irradiation-processed enamels yielded smaller amounts of extractives than the thermo-processed enamels. These enamels had been irradiated with 6 Mrad at 25°C in the presence of the food-simulating solvents.

Following the same procedures described in the above-mentioned study, additional extractive data were obtained for the epoxy phenolic enamel that was irradiated at 4.7–7.1 Mrad at 25 and −30°C in the presence of distilled water, 3% acetic acid, and n-heptane. The changes in the amount of extractives resulting from the irradiation treatment are shown in Table IX. In the case of the water and acetic acid extractives, there was no change in either the chloroform-soluble fractions or the chloroform-insoluble fractions. In the case of the n-heptane extractives, the amount of extractives decreased when the irradiation temperature was reduced from +25 to −30°C. Infrared spectra of the chloroform-soluble residues from the water and acetic acid extractives of the un-irradiated and irradiated enamel were identical to the chloroform-soluble residues from the solvent blanks. In other words, the epoxy phenolic enamel had no chloroform-soluble residue from the water and acetic acid extractives that could be attributed to the enamel. The n-heptane-soluble residue of the irradiated enamel was identical to the residue found in the unirradiated enamel, being a low molecular weight residue of the parent enamel.

Production Test. In a small-scale production test, tinplate containers with two commercially available enamels and two end-sealing compounds, which were selected from the results of this study, performed satisfactorily when packed with beef and ham. Beef was irradiated with 4.5–5.6 Mrad at 5, −30, and −90°C; ham was irradiated with 3–4 Mrad and 6–7.5 Mrad at −30°C. For this production test, beef and ham were packaged in round tinplate containers and ham in Pullman tinplate containers, frozen and refrigerated products were shipped 1,200 miles by truck, and were gamma irradiated at various doses and temperatures. Irradiated products were shipped 1,200 miles in a non-refrigerated truck and stored at selected temperatures and humidities. The integrity of the cans was evaluated after storage for 10 days, 3 months, and 6 months.

Table IX.   Change in Amount of Extractives After Irradiation with 4.7–7.1 Mrad

	Change in Extractives, mg/cm^2	
Solvent	25°C	−30°C
Water		
$CHCl_3$ soluble	0.00	0.00
$CHCl_3$ insoluble	0.00	0.00
Acetic acid		
$CHCl_3$ soluble	0.00	0.00
$CHCl_3$ insoluble	0.00	0.00
n-Heptane	0.002	0.001

The production test showed that the epoxy phenolic enamel was the preferred enamel for coating tinplate containers used in packaging irradiation-sterilized ham and beef. The preferred end-sealing compound for the same application was the blend of cured and uncured isobutylene–isoprene copolymer.

Conclusions

The evaluation of the components of the tinplate container showed that the preferred enamel for irradiation processing was the epoxy phenolic; the preferred end-sealing compound was the blend of cured and uncured isobutylene–isoprene copolymer. Component testing of tinplate and solder for possible changes in mechanical properties, microstructure, and corrosion resistance indicated that the radiation caused no tin rot, *i.e.*, conversion of tin from the beta, or silvery form, to the alpha, or powdery form. In the small-scale production test the tinplate container was reliable for packaging irradiation-sterilized beef and ham under adequate production conditions.

Literature Cited

1. Pratt, G. B., U.S. Army Natick Laboratories Contract Report **DA19-129-QM-968,** American Can Co., 1960.
2. Brillinger, J. H., Kalber, W. A., U.S. Army Natick Laboratories Contract Report **QMR&D63,** Dewey and Almy, 1958.
3. Pratt, G. B., U.S. Army Natick Laboratories Contract Report **DAAG17-68-C-0174,** American Can Co., 1970.
4. Holm, N. W., Jarrett, R. D., "Radiation Preservation of Foods," National Academy of Sciences–National Research Council, U.S.A., Publication **1273,** 1965.
5. Lewis, A. F., Gillham, J. K., *J. Appl. Polym. Sci.* (1963) **7,** 2293.
6. Hoare, W. E., Britton, S. C., Tin Research Institute (England), Publication **313,** 1964.
7. Ebben, G. J., Lawson, G. L., *J. Appl. Phys.* (1963) **34,** 1825.
8. Adams, F. W., "Analytical Problems Associated with Food Packaging Materials," Food Packaging Materials Section, *Int. Congr. Pure Appl. Chem. London* (1963) **19.**
9. Dole, M., *Radiat. Chem. Macromol.* (1973) **2,** 98.

RECEIVED April 12, 1974.

Compatibility of Aluminum for Food Packaging

M. A. JIMENEZ and E. H. KANE

Packaging Research Div., Reynolds Metals Co., Richmond, Va. 23219

Aluminum is one of the most common materials used for food packaging. Alloying elements are added to pure aluminum to improve its physical and chemical properties. Coating materials or plastic laminants applied to aluminum containers and flexible foil packages improve their end-use performance. The final selection of a package requires thorough compatibility testing with the specific product to be used. Test results on the interactions of various foods and beverages with aluminum containers are extensively discussed. The electrochemical action of foods wrapped with aluminum foil and then placed in contact with other metallic objects is reviewed. Aluminum and its salts have a harmless effect when ingested with foods that have been exposed to the metal.

The use of aluminum for food containers started in Europe more than half a century ago. Those containers, however, were made without considering the special properties of aluminum or the requirements of the food. Over the last two decades, the commercial applications of aluminum for packaging have significantly increased because of technological advances in metallurgy and container-making processes. In recent years, food products of many types have been sold in aluminum packaging or wrapped with aluminum foil. In 1972, the amount of aluminum (millions of pounds) used in the United States was as follows (*1*):

Metal cans	— 1,139	Caps and closures	— 60
Consumer foil	— 190	Composite cans	— 34
Flexible packaging	— 166	Industrial foil	— 28
Semi-rigid containers	— 165	Other	— 40

A large percentage of the metal cans was used by the beverage and food industries. Aluminum is now one of the major materials that people immediately consider for packaging possibilities. The popularity of aluminum stems from the fact that it is tasteless, odorless, nontoxic, light in weight, and a good barrier to the passage of gases, moisture, light, and grease. It has excellent thermal conductivity characteristics, and its strength and ductility increase as temperature decreases.

Alloying Elements

Most commercial uses of aluminum require special properties that the pure metal cannot provide. The addition of alloying elements imparts strength, improves formability characteristics, and influences corrosion resistance properties. The general effect of several alloying elements on the corrosion behavior of aluminum has been reported by Godard et al. (2) as follows:

Copper reduces the corrosion resistance of aluminum more than any other alloying element. It leads to a higher rate of general corrosion, a greater incidence of pitting, and, when added in small amounts (for example, 0.15%), a lower rate of pitting penetration.

Magnesium has a beneficial influence, and Al–Mg alloys have good corrosion resistance.

Manganese slightly increases corrosion resistance.

Silicon slightly decreases corrosion resistance. Its effect depends on its form and on its location in the microstructure of the alloy.

Chromium increases corrosion resistance in the usual amounts added to alloys (0.1–0.3%).

Table I.   Chemical Composition Limits of Wrought Aluminum

	Alloy Designation		
Element	1100	3003	3004
Silicon	} 1.0	0.6	0.30
Iron		0.7	0.7
Copper	0.05–0.20	0.05–0.20	0.25
Manganese	0.05	1.0–1.5	1.0–1.5
Magnesium	—	—	0.8–1.3
Chromium	—	—	—
Zinc	0.10	0.10	0.25
Titanium	—	—	—
Others, each	0.05	0.05	0.05
Others, total	0.15	0.15	0.15
Aluminum	99.00	remaining	remaining

ᵃ Composition in percent maximum unless shown as range

Zinc has only a small influence on corrosion resistance in most environments. It tends to reduce the resistance of alloys to acid media and to increase their resistance to alkalies.

Iron reduces corrosion resistance. It is probably the most common cause of pitting in aluminum alloys.

Titanium has little influence on corrosion resistance of aluminum alloys.

Other elements. A limited amount of information has been published on the effect of other elements, and the influence of many of them is still unknown.

Table I shows the chemical composition limits of various aluminum alloys presently used for packaging applications (3). In general, these alloys have good corrosion resistance with most foods. However, almost without exception, processed foods require inside enameled containers to maintain an acceptable shelf life (4, 5). Moreover, when flexible foil packages are used for thermally processed foods, the foil is laminated to plastic materials that protect it from direct contact with the food and also provide heat sealability as well as other physical characteristics (6, 7).

Aluminum–Food Compatibility Test Procedures

Although many packaged foods and beverages are systematically evaluated in commercial laboratories and although much experience has also been gained through the years, it is extremely difficult to predict the results of each individual food or beverage formulation because of the many possible combinations of coating materials and plastic laminants

Alloys Used for Fabricating Cans, Containers, and Consumer Foil [a, b]

Alloy Designation			
5050	5052	5182	8079
0.40	0.45 }	0.20	0.05–0.30
0.7		0.35	0.7–1.3
0.20	0.10	0.15	0.05
0.10	0.10	0.20–0.50	—
1.1–1.8	2.2–2.8	4.0–5.0	—
0.10	0.15–0.35	0.10	—
0.25	0.10	0.25	0.10
—	—	0.10	—
0.05	0.05	0.05	0.05
0.15	0.15	0.15	0.15
remaining	remaining	remaining	remaining

b (3)

with different aluminum alloys. Therefore, each new product has to be thoroughly tested to determine its compatibility with the specific package contemplated. The economics involved and the fabrication feasibility of the package must also be considered when making the selection.

The criteria used to establish the compatibility of aluminum with different products consisted of:

1. Determining changes in pH value and total acidity
2. Analyzing critical substances naturally present or intentionally added to foods
3. Analyzing traces of aluminum and other pertinent elements
4. Conducting sensory evaluations
5. Testing the vacuum of hermetically sealed containers
6. Analyzing headspace gases
7. Examining the condition of coating materials
8. Inspecting packages for evidence of corrosion.

Table II. Effect of Ascorbic Acid on the Color of Applesauce Stored at 75°F in Various Types of Cans[a]

Type of Can	Ascorbic Acid Added to Applesauce (ppm)	Mean "L" Values[b]	
		4 Months	15 Months
Single-coated aluminum	300	47.8	46.1
Single-coated aluminum	1,000	50.5	47.4
Double-coated aluminum	300	50.0	46.7
Double-coated aluminum	1,000	47.6	45.7
Tinplate (plain body, "F" ends)	none	51.1	49.9

[a] Adapted from (8)
[b] Obtained with a Hunter Color and Color difference Meter, Model D25

Aluminum–Food Compatibility Test Results

By classifying foods and beverages according to their acid content or pH value, it is possible to predict their compatibility with aluminum to some extent. However, the interactions between products and packages are also affected by natural substances present in some foods or intentionally added to some products, which inhibit or accelerate corrosion. For instance, natural pigments such as the anthocyanins and chlorophyll or added substances such as salt (sodium chloride), sugar, gelatin, synthetic dyes, nitrates, nitrites, and phosphates can be more significant than the effect of acidity alone. Although the following product categories are based on similarities in general chemical composition, the specific behavior of a product with respect to aluminum depends on the combined interactions of the various factors mentioned.

Table III.   Effect of Temperature on the Vacuum of Canned Juices Stored in Aluminum Cans[a]

Can Vacuum in Inches of Hg

		75°F		100°F	
Product	*2 Days*	*4 Mos.*	*12 Mos.*	*4 Mos.*	*8 Mos.*
Uncoated cans					
Orange juice	swell	—	—	—	—
Pineapple juice	swell	—	—	—	—
Tomato juice	8	4	swell	2	swell
Vegetable juice blend	15	11	swell	swell	—
Single coated cans					
Orange juice	23	21	21	20	20
Pineapple juice	23	22	21	20	20
Tomato juice	23	21	21	20	20
Vegetable juice blend	22	20	20	18	18

[a] Adapted from (8).

Fruit and Vegetable Products. Studies on several fruits packed in light syrup (8) showed that they attacked the uncoated aluminum. The accumulation of hydrogen gas formed by corrosion caused the cans to swell. An internal coating significantly decreased the attack on the containers, but some color changes occurred during processing and storage. The degree of discoloration was similar to that sometimes observed in glass-packed fruits. As shown in Table II, applesauce darkening in coated aluminum cans was almost totally inhibited by adding ascorbic acid (Vitamin C) to the product. This problem does not exist with tinplate cans because of the bleaching effect of tin on applesauce. Jam and jellies are much less corrosive to aluminum than canned fruits in syrup because of the protective action of their large sugar content (9) and the high viscosity of these products (10).

Fruit and vegetable juices packed with 21–26 in. of vacuum and stored in uncoated aluminum cans caused severe corrosion as shown in Table III. The corrosion rate brought about by the juices depends more on the nature of the organic acid present and the buffering capacity of the juice than on the total titratable acidity (11). The use of coated aluminum containers considerably minimized corrosion problems. Product control under extended storage conditions may be achieved by using specific chemical additives. However, more work is needed in this area before final conclusions can be reached.

Among the vegetables tested, peas and corn with pH values of 6.0 and 6.1, respectively, caused the least corrosion to plain aluminum cans. In fact, Lopez and Jimenez reported (8) that whole kernel corn retained a brighter yellow color in uncoated aluminum cans than in tinplate or aluminum containers that had an internal coating. Green beans, spinach, beets, and asparagus with pH values between 5.3 and 5.5 also corroded the uncoated aluminum cans causing hydrogen swells. It is likely that the salt (sodium chloride) added to these products accelerated the attack on the metal. According to Jakobsen and Mathiesen (12), the corrosive action of spinach has been attributed to the high level of water soluble oxalic acid occasionally found in this vegetable. This effect can be suppressed by adding a small amount of calcium chloride to the spinach. The calcium oxalate that is formed has no corrosive action.

Table IV. Internal Coatings Used on Aluminum Cans for Fruit and Vegetable Products[a]

Product	pH	Type of Internal Coating
Apricots	3.7	phenolic, phenolic oleoresinous
Cherries	3.8	phenolic
Pimientos	4.3	modified epoxy
Tomatoes	4.3	phenolic, phenolic oleoresinous
Bean in sauce	5.2	phenolic
Green beans	5.3	phenolic
Spinach	5.4	phenolic
Asparagus	5.5	phenolic
Peas	6.0	phenolic

[a] Adapted from Ref. 8.

Table IV shows various types of organic coatings used in the interior of aluminum cans tested with different fruit and vegetable products. These coatings resisted physical stresses during heat processing, but could not prevent changes in flavor and color of the products. Subsequent work using inert coatings and films, such as modified polypropylene and nylon 11, combined with flexible and semirigid aluminum structures has shown that the color and flavor of vegetables in aluminum packaging is equal to and often better than in the familiar canned foodstuff.

Meat Products. Laboratory investigations as well as practical experience have shown that aluminum containers do not cause the objectionable darkening of meat that occurs with other metals. Wunsche (13) found that luncheon meat stored in lacquered aluminum cans retained its normal color after more than one year, while a slight surface discoloration was observed on the same product packed in lacquered tinplate cans.

Foods such as meat, fish, and some vegetables contain sulfur-bearing amino acids that form volatile sulfur compounds during processing and storage. When these compounds react with iron, a black precipitate forms on the container and in most instances darkens the food. A small piece of aluminum welded to the tinplate can has been used to prevent container corrosion and sulfide staining in commercially canned hams. In this case, the aluminum acts as a sacrificial anode and stops the reaction with tin and iron that otherwise could occur at the small exposed tinplate areas (*14*).

The addition of gelatin to meat products, especially those highly salted or containing other corrosive ingredients, reduces the attack on the metal. The shelf life of canned meats is also significantly extended by effectively removing the air from the container headspace.

Fish and Shellfish. Sardines in oil and also in tomato sauce and mustard sauce are packed commercially in enameled aluminum cans. However, tomato and mustard sauces are corrosive products that can attack metal containers. Sardines prepared in these sauces should not exceed 3.0% total acidity, expressed as acetic acid. Otherwise, the presently used interior can enamels will not protect the food sufficiently to prevent chemical reactions with the metal.

When packaging lobster in tinplate cans, parchment paper is used to prevent product discoloration caused by the iron present. This is not required with aluminum cans. Studies conducted with canned shrimp have shown that the desirable pinkish cast and color bands normally associated with recently cooked shrimp are bleached when in contact with various metals. According to Gotsch *et al.* (*15*), this reaction causes shrimp to turn gray and to develop a hydrogen sulfide-like odor in storage. Landgraf (*16*) has reported that these problems can be reduced significantly by lowering the pH of the product to 6.0–6.4 by adding citric acid to the brine. Thompson and Waters (*17*) have recommended the use of either lemon juice or citric acid for this purpose.

Milk Products. The packaging of milk products in aluminum containers usually requires a protective coating. This is especially true with unsweetened evaporated milk since it does not have the added sugar content of sweetened condensed milk. Careful selection of the container coating is very important since milk is extremely sensitive to flavor and color changes during heat processing and storage. With aluminum containers there is no problem of possible contamination with lead that can develop when these products are packed in soldered tinplate cans.

Flavor and odor preservation is a critical requirement with butter, which also needs refrigeration to avoid rancidification. Light and oxygen promote photochemical oxidation of this product. Aluminum foil provides opacity and has excellent barrier properties (*18*). The material generally used to package butter is thin gauge, aluminum foil laminated paper. Gourmet types of butter flavored with garlic or spices are also marketed in formed foil containers. For convenience, small portion control units in formed aluminum foil are often used in butter packaging. Although not a dairy product, soft margarine can also be packaged in sealed aluminum containers when maximum shelf life is desired. Aluminum cans are also widely used for packaging a complete variety of shelf-stable puddings.

Any material used for packaging natural cheeses must prevent moisture loss, maintain good product appearance, protect against microorganisms, and act as an oxygen barrier. Aluminum foil laminates provide this type of protection for cheese. Cream cheese is packed in lami-

Table V. Effect of Iron and Aluminum on the Quality of Canned Beer[a]

Storage Time (months)	Iron[b]	Aluminum[b]	Flavor	Clarity	Color
Double-Enameled Tinplate Cans					
0	.18	.08	normal	excellent	light
1	.28	.07	normal	excellent	light
3	.49	.10	oxidized	good	slightly dark
6	.84	.10	extremely oxidized	fair	slightly dark
Single-Enameled Aluminum Cans					
0	.07	.27	normal	excellent	light
1	.10	.32	normal	excellent	light
3	.09	.40	slight oxidized	very good	light
6	.08	.45	slightly oxidized	good	slightly dark

[a] At 77°F, (19).
[b] In ppm

nated foil structures such as foil–paper–foil that have excellent folding characteristics coupled with protection. Port du Salut cheese is wrapped in aluminum foil, and Roquefort is packaged in an aluminum foil laminate. Hermetically sealed, formed aluminum containers are also popular for packaging many cheese spreads.

Beer. Laboratory results obtained by Jimenez and Gauldin (19) as well as commercial experience have shown that beer in aluminum cans is superior in flavor, color, and clarity to beer packed in tinplate cans. Table V summarizes the effect of aluminum and iron on beer stored for six months in the two types of metal containers. Aluminum ends used in conjunction with tinplate or tin-free-steel can bodies increases the shelf life of beer because of the sacrificial role of aluminum in retarding the iron pickup rate. All aluminum cans with vinyl epoxy or clear epoxy coatings are particularly favorable for use with beer.

Carbonated Soft Drinks. Although there is not enough data available to establish maximum levels of dissolved aluminum for each soft drink formulation, Lemelin (20) has reported that cola and lemon-based drinks containing 5–10 ppm aluminum showed no significant flavor deterioration after six months at 78°F. A relatively high amount of dissolved aluminum will not adversely affect the flavor of soft drinks.

Metal exposure is a critical factor in color retention of most azo dye formulations. However, improved coating techniques have made it possible to attain the desired shelf life of products colored with azo dyes.

Corrosion has been encountered infrequently to date and has been a surface type, as opposed to pitting corrosion that can result in perforations. Entrapped air in the beverage or in the can's headspace increases the corrosive action of the product according to Koehler *et al.* (21). As with beer and other canned foods, aluminum ends provide electrochemical protection when combined with tinplate or tin-free-steel can bodies. The level of iron pickup is reduced while the amount of aluminum dissolved in soft drinks increases without detrimental effect. Aluminum containers with vinyl epoxy and vinyl organosol coatings are compatible with carbonated soft drinks.

Strong Alcoholic Beverages. Products such as whiskey, cognac, brandy, etc. cause undesired reactions with unprotected aluminum. The attack causes pitting corrosion and formation of a floculent precipitate of aluminum hydroxide while the beverage itself becomes discolored, and the flavor is also affected (22). The action of liqueurs is not so

Table VI.   Effect of Air and Aluminum on SO_2 Retention of Wines Stored in Double-Enameled Aluminum Cans[a]

Storage Time at 80°F (months)	Nitrogen Gas Flow Closure		Atmospheric Closure	
	Aluminum Pickup (ppm)	SO_2 Retention (%)	Aluminum Pickup (ppm)	SO_2 Retention (%)
Dessert Wines				
3	<1	97	<1	54
6	<1	91	<1	49
12	<1	81	1.6	47
Table Wines				
3	<1	96	<1	58
6	<1	89	1.4	49
12	<1	88	2.1	52

[a] (23).

pronounced because of the inhibiting action of the sugar present. Hotchner and Schild (23) have demonstrated that double enameled aluminum cans have been successfully used for packaging premixed alcoholic cocktails, but it is extremely important that the air be effectively removed from the container. These products must also contain no more than 0.5 ppm dissolved copper.

Wines. Wines are very sensitive to oxygen and certain metals with respect to color, cloudiness, flavor, and bouquet. The amount of sulfur dioxide added to wines to prevent microbial spoilage has to be retained to maintain its quality. In attempting to package wine in metal containers, the reducing action of the metal on the sulfur dioxide must be prevented. This can be done by applying a protective coating to the container. It is also critical to flush out the air with nitrogen gas (23). If this is effectively accomplished, it is feasible to package wine in aluminum cans. As shown in Table VI, the level of dissolved aluminum must be restricted to less than one ppm.

Tea Beverages. Uncoated aluminum is not compatible with tea beverages since direct contact causes discoloration and undesired flavor changes during storage. When the water contains dissolved iron, it can react with the tannin in the tea and cause darkening in the presence of air. Dissolved air in the beverage or in the headspace of the can should be

Table VII. Aluminum Pickup by a Tea Beverage Stored in Single-Coated Aluminum Cans with Easy-Open End[a]

Storage Time (months)	Aluminum pickup (ppm)	
	Can Stored Upright at 78°F	Can Stored Inverted at 100°F
1	1.4	1.8
3	2.6	4.8
6	2.8	4.8
12	3.2	—

[a] (24)

flushed with nitrogen gas. In the absence of oxygen, the iron stays in the divalent or incompletely oxidized state and forms a colorless iron–tannin compound. When the air is not effectively removed, the iron is oxidized to the trivalent state, forming a colored iron–tannin compound that darkens the beverage. A sweetened and acidified tea beverage formulation packed in single enameled aluminum cans flushed with nitrogen gas showed no significant discoloration or flavor changes after 12 months in storage. Table VII illustrates the rate of aluminum pickup by a tea beverage stored in aluminum cans kept at two different temperatures for one year (24).

. **Frozen Foods.** Corrosion caused by the reaction of foods with aluminum containers is unusual if the products are handled and stored at 0°F or lower. However, the inevitable bad handling of frozen foods during commercial distribution causes undesirable thawing. In this condition, not only does the food deteriorate, but it can also attack the container. Such unwanted reactions can be effectively controlled by using coated aluminum containers. Since aluminum is highly compatible with frozen fruits and citrus juices, it has been used extensively as a liner for fiberboard composite cans, as complete aluminum cans, or as ends in combination with steel can bodies in the frozen food industry.

Types of Aluminum Food Packaging

Aluminum Foil. Studies of various foods wrapped in aluminum foil show that food products to which aluminum offers only fair resistance cause little or no corrosion when the foil is in contact with a nonmetallic object (glass, plastic, ceramic, etc.) The reactions, when found, are essentially chemical, and the effect on the foil is insignificant. However, when the same foods are wrapped or covered with foil that is in contact with another metallic object (steel, tinplate, silver, etc.), an electrochemical or galvanic reaction occurs with aluminum acting as the sacrificial anode. In such cases, there is pitting corrosion of the foil, and the severity of the attack depends primarily on the food composition and the exposure time and temperature. Results obtained with various foods cov-

Table VIII.   Interaction of Aluminum Foil in Contact with Foods in Different Containers[a]

| | | Corrosion of Foil in Contact with Food and Container at 36°F | | | |
Food Product	Storage Time	Glass	Aluminum	Steel[b]	Silver[c]
Spaghetti (meat sauce)	7 days	none	none	pitting	—
Lasagna	24 hours	none	none	pitting	—
Honey-glazed ham	16 hours	none	none	—	pitting
Open-faced sandwiches	24 hours	none	none	—	pitting
Potato salad	24 hours	none	none	pitting	pitting
Jellied fruit salad	24 hours	none	none	pitting	pitting
Pumpkin pie	18 hours	none	none	pitting[d]	—

[a] (25).
[b] Stainless steel tray
[c] Silver-plated tray
[d] Noticeable after two hours

ered with foil and placed in contact with other objects are shown in Table VIII (25).

Flexible and Semirigid Aluminum Packaging. Most of the discussion on the use of aluminum for food packaging has centered on rigid cans. However, aluminum is widely used in flexible specifications and semirigid containers for the protective packaging of a tremendous variety of food products. For example, laminated aluminum pouches are extremely popular for packaging moisture-sensitive products such as snack foods, bakery items, drink mixes, salad dressings, confectionery products, cereals, cake mixes, etc. Liquid and semiliquid products are also packed in flexible aluminum structures as well as in formed aluminum containers.

As a result of extensive development and testing by thermoprocessing or aseptic techniques, the use of flexible, laminated aluminum pouches and formed aluminum containers for shelf-stable foods is nearing commercial reality. The increasing use of aluminum for food packaging has been made possible by successfully combining it with specialized plastics, papers, adhesives, and coatings. In many applications, aesthetic as well as protective characteristics are also provided.

Aluminum and Health

Aluminum has a long history of safe usage in connection with food and food packaging. Moreover, aluminum is generally recognized as safe (GRAS) by the U.S. Food and Drug Administration. In spite of this, statements appear from time to time alleging that the action of foods on aluminum forms toxic substances that cause diseases, destroy vitamins, etc. These allegations are absolutely without factual basis.

Results published and statements issued by recognized health specialists (26–28) confirm that aluminum and its salts are innocuous in the quantities ingested with foods and beverages that have been exposed to the metal. Juniere and Sigwalt (29) reported that aluminum is present in small amounts in a large number of plant and animal products. This is not surprising since aluminum is the most abundant and widely distributed of all metals. Practical tests have also shown that quantities of aluminum far in excess of those that may be dissolved by prolonged cooking of foods can be ingested without harm. The Department of Preventive Medicine and Industrial Health of the College of Medicine of the University of Cincinnati has published the following statement (30):

> There is no reason for concern on the part of the public or of the producer and distributor of aluminum products about hazards to human health derived from well established and extensive current uses of such products. Nor need there be concern over the more extended uses which would seem to be in the offing.

Summary

Aluminum is one of the most common materials used for food packaging applications. The interactions between aluminum and foods or beverages require, almost without exception, that the metal be protected from direct contact with the product to increase its corrosion resistance and to prevent undesired reactions that may impair the product quality. Suitable coatings or plastic laminants significantly minimize such interactions by acting as a protective barrier. However, the ultimate selection of a package requires thorough testing to determine its compatibility with the specific product contemplated. Foods wrapped with foil should not be in contact with other metallic objects because electrochemical reactions that cause severe corrosion could occur. Medical authorities have confirmed that aluminum or its salts are innocuous in the quantities ingested with foods that have been exposed to the metal. Therefore, health hazards should be of no concern with the many forms of aluminum packaging used with foods and beverages.

Acknowledgment

The authors express their appreciation to W. Page Andrews, General Director of Packaging Research, Reynolds Metals Co., for reviewing the manuscript.

Literature Cited

1. *Mod. Packag.* (1973) **46**, (7), 10.
2. Godard, H. P., Jepson, W. B., Bothwell, M. R., Kane, R. L., "The Corrosion of Light Metals," John Wiley & Sons, Inc., New York, 1967.
3. "Registration Record of Aluminum Association Alloy Designations and Chemical Composition Limits for Wrought Aluminum Alloys," The Aluminum Association, New York, 1973.
4. Adam, W. B., *Food Preserv. Quart.* (1951) **11**, (3–4), 35.
5. Taranger, A., *Light Metals* (1956) **19**, 387.
6. Dirdjokusumo, S., Luh, B. S., *Food Technol.* (1965) **19**, 1144.
7. Goldfarb, P. L., Packaging Institute, Publication No. **T7309**, 1973.
8. Lopez, A., Jimenez, M. A., *Food Technol.* (1969) **23**, 1200.
9. Beal, G. D., Unangst, R. B., Wigman, H. B., Cox, G. J., *Ind. Eng. Chem.* (1932) **24**, 405.
10. Teschner, F., *Light Metals* (1939) **2**, 201.
11. Datta, N. C., *Chem. Abstr.* (1936) **30**, 525.
12. Jakobsen, F., Mathiesen, E., "Corrosion of Containers for Foods," p. 93, I Kommisjon Hos Jacob Dybwad, Oslo, 1946.
13. Wunsche, G., *Verpack. Rundsch.* (1968) **19**, (9), 1074.
14. McKernan, B. J., Davis, R. B., Fox, J. F., Johnson, O. C., *Food Technol.* (1957) **11**, 652.
15. Gotsch, L. P., Eike, E. F., Brighton, K. W., *Proc. Ann. Conv. Nat. Canners Ass.*, Information Letter No. **1720**, 99.
16. Landgraf, R. G., Jr., *Food Technol.* (1956) **10**, 607.
17. Thompson, M. H., Waters, M. E., *Commer. Fish. Rev.* (1960) **22**, (8), 1.

18. Sacharow, S., Griffin, R. C., Jr., "Food Packaging," pp. 157–165, Avi, Westport, 1969.
19. Jimenez, M. A., Gauldin, E., *Proc. Amer. Soc. Brewing Chem.* (1964) 233.
20. Lemelin, D. R., *Proc. Soc. Soft Drink Technol.* (1969) 85.
21. Koehler, E. L., Daly, Jr., J. J., Francis, H. T., Johnson, H. T., *Corrosion* (1959) **15**, (9), 45.
22. Bryan, J. M., Low Temperature Research Station, *Spec. Rept.* **50**, Cambridge, England, 1948.
23. Hotchner, S. J., Schild, C. W., *Food Technol.* (1969) **23**, 773.
24. Reynolds Metals Co., Internal Report, 1973.
25. Reynolds Metals Co., Internal Report, 1972.
26. Bordas, F. B., Report Council Public Hygiene, France, 1929.
27. Mellon Institute, *Ind. Res., Bibliog. Ser., Bull.* **3** (1933).
28. Wuhrer, J., *Metals Alloys* (1939) **10**, 376.
29. Juniere, P., Sigwalt, M., "Aluminum, Its Application in the Chemical and Food Industries," Chemical Publish. Co., New York, 1964.
30. Univ. of Cincinnati, College of Medicine, Department of Preventative Medicine and Industrial Health, "Aluminum and Health," Dudley-Anderson-Yutzy, New York, 1957.

RECEIVED October 1, 1973.

Packaging Food Products in Plastic Containers

CHARLES A. SPEAS

Hedwin Corp., 1600 Roland Heights Ave., Baltimore, Md. 21211

Plastics packaging and contained food products are chemically related in four distinct ways. This relationship is based largely on the permeation property of the plastic material. Direct chemical reaction between plastic and product is seldom a problem when inert plastics such as polyethylene are used. However, polyethylene can transmit minute amounts of product to the outside. This paper examines the effect of permeation through the plastic wall and the direct effects of the plastic on the food product. Specific food packaging applications and methods of testing are discussed.

The chemical relationship between plastic packaging and food products is influenced by the characteristics of the plastics materials that are used. This paper includes test results and observations of field performance made by a container manufacturer while designing and developing 1–55 gallon polyethylene containers for the bulk shipment and storage of liquid food products. However, most of the theory, principles, and test methods discussed are applicable to all food packaging, regardless of container size or type of plastic.

Plastic as a Food Packaging Material

The mutual chemistry of plastic containers and food products must be considered for any proposed application. There is continuous physical and chemical activity at the interface between the food product and the container. The type and extent of this activity determines whether or not the plastic container can successfully hold and protect the food product. However, the U.S. Food and Drug Administration and the American public are increasingly suspicious of all plastics, particularly the halogenated compounds. The recent ban (April 1973) on poly(vinyl chloride) bottles for alcoholic beverages will hopefully be lifted, but only if it can be proved that vinyl chloride monomer is not toxic or carcinogenic.

Such an experience with one plastic, PVC, makes it doubly important to carefully examine any plastic to be used with a food product. The basic question to be answered is "Does the plastic container provide adequate protection to the food product during the entire life cycle of the container?" Adequate protection of a food product in a polyethylene container implies that there is no undesirable change in the chemical content of the food during storage in the container. Thus, our study is concerned with the ways in which food products can change when stored in polyethylene containers.

Polyethylene, because of its inertness and physical properties, is nearly ideal as a packaging material for food products. It is inert to most chemicals and, except for overheating (which can oxidize it) or irradiation (which can cross link the molecules), the chemical composition of polyethylene is difficult to change. There is one important property which it does lack—permeation resistance.

The Monsanto Company developed the Permachor method to predict permeation through polyethylene and other polymers. The prediction is

Figure 1.   The container must block   Figure 2.   The container must pre-
        entry of outside gases.   vent product loss.

based on the size, shape, and polarity of the permeating molecule and has been used extensively by the plastic container industry. However, as the paper by Salame and Temple of Monsanto indicates (*1*), there is no substitute for actual testing, especially with food products.

Interactions Between Container and Product

Adequate protection of a food product by a plastic container should involve careful consideration of the following four questions:

1. Does the container adequately block the entry of outside gases and light (especially *UV*) (Figure 1)?
2. Does the container keep the product and its essential components from escaping (Figure 2)?
3. Does the container wall absorb a significant amount of any essential ingredient (Figure 3)?
4. Does the polyethylene contribute undesirable material to the food product (Figure 4)?

Selection of a polyethylene resin with no additives or with FDA qualified additives usually insures good direct mutual chemistry between the plastic and the food product, and many food products can be shipped

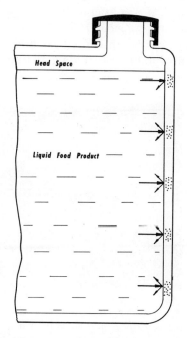
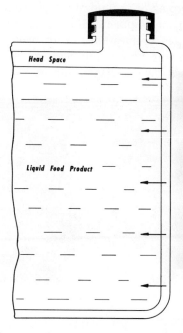

Figure 3.   The container wall must not absorb essential ingredients.

Figure 4.   The container wall must not add components to the product.

and stored in polyethylene containers with minimal pretesting. This re-sults from the excellent physical properties and the chemical inertness of this plastic. Exceptions caused by varying temperatures and storage times require caution and study. However, the most common limitation is permeation. A product may be safely held for months under refriger-ated conditions whereas a week in the same container at elevated tem-peratures might be unacceptable. Thus storage conditions must be considered for every application. Damage from external light, such as UV, could be a problem with certain food products. However, the com-bination of the opaque plastic and the corrugated fiberboard shipping carton is sufficient to negate the effect of UV radiation.

Oxygen Permeation

Perhaps the best way to examine the relationship between poly-ethylene containers and liquid food products is to examine several applications which posed problems related to one or more of the above four attributes.

Figure 5 shows a molded 1-gallon, low density polyethylene liner in a corrugated carton. Since this container had an excellent history of shipping 85% food grade phosphoric acid, it was examined in a 4-gallon size for shipment and storage of a beverage concentrate. The inward oxygen permeation rate of over 100 cc per month caused no problems with the phosphoric acid but was unacceptable for holding the citrus-based beverage concentrate. Taste tests conducted before and after a

Figure 6. Partly assembled, 5-gallon low density polyethylene liner in a cor-rugated carton. Liner is coated with poly(vinylidene chloride) (Saran) to block oxygen entry and prevent loss of aroma used for cola concentrate.

Figure 5. One-gallon, low density polyethylene liner in a corrugated car-ton. Used with various chemicals, in-cluding food grade phosphoric acid.

prescribed storage period showed a noticeable effect on taste apparently resulting from oxygen permeation inward and aroma loss outward. A poly(vinylidene chloride) coating was applied to the exterior surface of the larger container (shown partly assembled in Figure 6). This reduced the permeation rate by 10:1, and taste tests results were comparable with storage of the concentrate in a glass container.

Taste testing is normally conducted before and after a holding test under normal warehousing conditions. However, there are some interesting variations applied to the holding conditions. For example, the more conservative beverage concentrate manufacturers also test by holding the filled containers alongside open cans of such odorous products as kerosine. They have found that exterior coatings are needed to provide

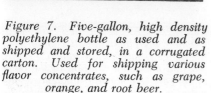

Figure 7. Five-gallon, high density polyethylene bottle as used and as shipped and stored, in a corrugated carton. Used for shipping various flavor concentrates, such as grape, orange, and root beer.

Figure 8. One-gallon, high density polyethylene bottle used for cooking oil. A special coating on the exterior preserves freshness by excluding oxygen.

the necessary protection in this situation. In another test series, two uncoated, low density polyethylene containers were stored side by side, one with grape concentrate and one with root beer concentrate. Taste tests showed an off taste in each caused by permeation inward by the other.

One corrective measure which is often employed when low density polyethylene has inadequate barrier properties is the use of high density instead of low density polyethylene. With this modification, there is a permeation improvement of approximately 3:1. Figure 7 shows a 5-gallon, density polyethylene bottle (shipped in a corrugated overwrap n) used by several beverage concentrate manufacturers for a wide variety of flavors including cola, orange, grape, and root beer.

Additional protection can be applied to the high density polyethylene by Saran or epoxy exterior coating. For example, Figure 8 shows a 1-gallon bottle blow molded of high density polyethylene. The container is currently being tested for cooking oil. Most cooking oil in the U.S. is sold in glass bottles, and the trend toward using plastic containers has not been strong. This would be surprising to European housewives, who are accustomed to buying cooking oil in bottles blow molded of poly(vinyl chloride). Cooking oil sold in such a container is very well protected because of the excellent oxygen permeation resistance of PVC. However, PVC is not as well accepted by the regulatory bodies and environmental groups in the U.S. as it is in Europe. Without some protection against oxygen in addition to that afforded by the high density polyethylene, the cooking oil manufacturer does not consider this 1-gallon bottle a satisfactory package. For this reason a Saran (poly(vinylidene chloride)) coating is applied to the exterior of the bottle which improves the permeation resistance of the uncoated high density bottle approximately ten-fold. This falls well within the time, temperature, and other parameters imposed by the cooking oil manufacturer's marketing conditions.

Figure 9. A coated and uncoated 1-gallon, high density bottle under test for oxygen permeation inward. Copper and ammonia solution in the container are chemically influenced by entering oxygen, which provides a quantitative basis for the measurement.

Considerable laboratory and field testing was required before even market test quantities of this coated 1-gallon bottle could be considered. In addition to extensive physical performance testing, oxygen permeation testing was required. The conventional method used for film and film laminate measurements is the gas transmission cell test (ASTM 1434). This test uses a manometric or volumetric method to measure gas flow rate through a sample disc several inches in diameter as a function of time, temperature, etc. More realistic and meaningful quantitative oxygen entry measurements are obtained by copper–ammonia analysis (2). For this test, the actual container is used instead of a sheet section of the sidewall. The bottle is filled with copper turnings and strong ammonia solution, as shown in Figure 9. During storage, oxygen which enters the container reacts with the copper to form copper–ammonia complexes. Colorimetric analysis determines the cupric ion concentration in the solution which is proportional to the oxygen concentration.

Product Loss

Because essential components such as flavor and odor constituents may be lost, a test program will insure that the proposed container does not lose enough of the flavor–odor components to significantly weaken or change the beverage taste. The most common evaluation method is the before and after taste test. This test is almost always used in arriving at a final verdict, but it may be supplemented by oxygen permeation tests (such as the copper–ammonia test previously described), sniff tests (determining with the nose the rate of odor loss from one container versus another), or instrument analysis of loss rate, such as by gas–liquid chromatography.

Gas–liquid chromatography testing (*3*) was used to determine permeation loss by comparing two special plastic containers which are small models of an experimental 5-gallon container designed to hold flavor concentrates (Figure 10). The plastic wall of the container was designed to provide close to 100% flavor loss protection, as well as an absolute barrier to entry of gases. The high barrier quality resulted from a layer

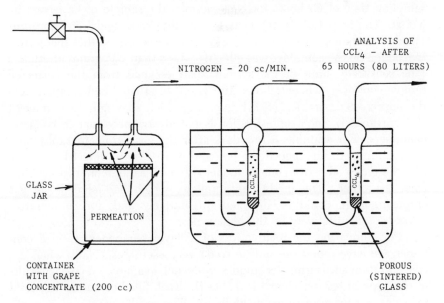

Figure 10.   Headspace analysis of poly foil pouches

of 0.00033 inch aluminum foil in the total laminate of paper, polyethylene, and aluminum (4). Each of the containers contained 200 ml of grape concentrate. One container had a known manufacturing defect, which could permeate but not leak liquid. The second sample was identical, except that the defect was corrected. Since the concentrate was grape, we were looking for grape aroma loss, specifically methyl anthranilate, which is the volatile constituent that contributes the grape taste and aroma (Table I).

Table I.   Volatile Constituents of Concord Grapes (5)

Constituent	(mg/ml/essence)
Ethanol	35.0
Methanol	1.5
Ethyl acetate	3.5
Methyl acetate	0.15
Acetone	0.3
Acetaldehyde	0.03
→Methyl anthranilate	0.033←
Acetic acid	
Unknown, CHCL$_3$-soluble	

Because we were unable to identify the methyl anthranilate component within the sensitivity of the equipment used for these tests, we resorted to an examination of the ethanol loss (Figure 11). After a sampling time of 65 hours for each sample, the sample of CCl$_4$ was injected. The amplitude of the peak at 1 minute 58 seconds retention time (peak for ethanol) was examined. The two tests proved that the corrective action on the container was effective. Less than 10^{-6} grams of ethanol (the sensitivity limit of the system) had escaped from the corrected sample container, whereas 1.5×10^{-4} grams of ethanol had escaped from the uncorrected sample. (See Figure 11, which shows the two traces.)

Fortunately, such sophistication is not always necessary to compare containers. For example, a few days after the gas chromatography tests were completed, we opened the jars and were able to detect by sniffing that the grape odor from one container was noticeably more distinct than from the other. The methyl anthranilate, which had eluded the gas chromatograph, could be detected qualitatively and, for comparative purposes, quantitatively by the human nose.

For comparing the relative loss of a flavor component from a container, we have found the sniff test (6) very useful, especially when gas concentration measuring techniques were not available. Typical results of this type of test are shown in Table II. Each filled container was held in a glass jar for approximately 48 hours. The results are stated in qualitative, subjective terms such as "slight," "strong," or "undetectable." The

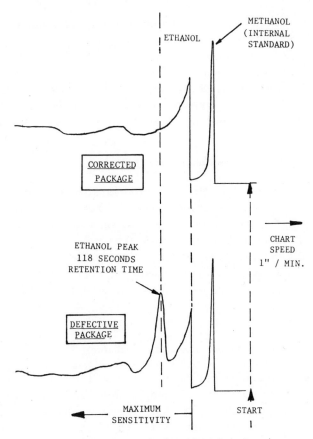

Figure 11.   Analysis for ethanol

Table II.   Aroma Loss Test Results[a]

Item Tested	After 2 Days	After 2 Weeks
Garlic concentrate		
Glass bottle	slight	moderate
Polyethylene film pouch	strong	very strong
Poly/foil/poly pouch	undetectable	very slight
Vanilla extract		
Glass bottle	very slight	moderate
Polyethylene film pouch	moderate	strong
Poly/foil/poly pouch	undetectable	very slight
Grape concentrate		
Glass bottle	slight	moderate
Polyethylene film pouch	moderate	very strong
Poly/foil/poly pouch	undetectable	very slight

[a] Using sniff test method with container enclosed in poly/foil/poly pouch with 38 mm screw cap.

test is particularly useful in comparing two types of container construction. The new container construction was, in each case, better than glass. The odor loss from each of the glass containers was, evidently, through the screw cap liner (7). The experimental containers were heat-sealed and had no screw cap openings. Although sniff tests are not quantitative and do not approach the accuracy of the GLC technique, they are more practical and more useful to the container designer.

Absorption of Product by the Container Wall

If a change occurs in the food product after storage in a plastic container, some part of the change could be caused by absorption in the container wall. The important components such as flavor oils or emulsifiers exist in relatively small quantities. The type and thickness of the polyethylene container can influence this variable. If the before and after taste test shows no difference between storage in the plastic container and storage in glass, absorption in the wall is considered insignificant.

Contribution to the Product from the Container Wall

The plastic container can influence the food product by direct contribution from the plastic. For example, milk and water, which have practically no aroma, cannot mask the very faint odor which may come from certain polyethylene formulations. The source of the faint odor in polyethylene, may be one of the following (8, 9):

1. Formation of carbonyl groups which may occur when polyethylene is overheated, such as in an extrusion coating operation. Here a proper balance between extrudate temperature and exposure time to air is required.

2. Residual catalyst (expressed as ash content). The actual residues depend on the manufacturing process used and on the characteristics of the polymer. The so-called new generation HD–PE processes such as the Solvay process use superactive catalysts which produce polymers with a low ash content and, hence, low or negligible odor. Some narrow MWD (Molecular Weight Distribution) resins also have lower catalyst residues than their wide MWD counterparts.

3. Antioxidant additive in the polyethylene resin. While such an additive can prevent oxidation, and thus odor, it also can contribute directly to the odor. If an antioxidant is needed, it must be FRA approved, should have a high melting point, and should be used at a minimum level consistent with the extrusion process. Catalyst residues and antioxidants present in polyethylene sometimes interact to form odorous products.

4. Low molecular weight fractions can be detected by smelling the inside of almost any freshly made polyethylene container. The amount varies with the specific polyethylene resin and the type of processing that have been used.

In some cases, the odor from one or more of the above sources may be strong enough to be detected on opening the container. The odor can exist in the head space and yet not be dissolved in the liquid in significant amounts. For most polyethylene containers designed to hold liquid food products, pouring the liquid into another container or dispensing it through a faucet below the liquid level of the container will eliminate odors from the above sources. There are exceptions, however. One is the blow molded, form-fill polyethylene bottles which are currently being made in Europe in sizes and types from which the user may drink directly. The container is blow molded with the liquid (instead of with air). Beverage and polyethylene are fed into the machine and molded, filled, and sealed bottles emerge. The odor produced from the polyethylene is captured and held. The problem has been solved by selecting a special resin type and using small quantities of a high melting point antioxidant.

Other Sources of Odor

On occasion the polyethylene has been blamed for odors caused by other parts of the container design such as screw cap liners, printing inks on the bottle exterior, adhesives used to laminate polyethylene to paper, and the fiberboard or the lining of material used in the separate exterior overpack in which a polyethylene container is held. Careful selection of all of these related materials is always advisable with liquid food products.

The proper combination of plastic and product will result in a good product with good mutual chemistry. With polyethylene and many food products, there are no problems to be solved. With other proposed combinations, a cooperative effort between the food product manufacturer, the container producer, and the resin supplier is needed to produce a high quality product.

Literature Cited

1. Salame, M., "The Prediction of Liquid Permeation in Polyethylene and Related Polymers," *SPE Trans.* (1961) **1**, (4).
2. Calvano, N. J., "O_2 Permeation of Plastic Containers," *Mod. Packag.* (November 1968).
3. Karasek, F. W., "Liquid Chromatography—HPLC," *Res. Develop.* (June 1973).
4. Pflug, I. J., Bock, J. H., Long, F. E., "Sterilization of Food in Flexible Packages," *Food Eng.* (1963) **17**, (9).

5. Webb, A. D., "Present Knowledge of Grape and Wine Flavors," *Food Technol.* (1962) **16**, (44).
6. Fontenot, J. L., "Kraft Board Odor Evaluation by Gas Chromatography and an Odor Judging Panel," *J. Ass. Pap. Pulp Ind.* (1972) **55**, (4).
7. Speas, C. A., "Closure Performance Requirements for High Barrier Plastics Liners for Fiberboard Cartons," *Proc. PIRA/IAPRI Int. Conf. Packag. Technol.*, London (March 27, 1972).
8. Waghorn, P. E., "The Control of Oxidation and Odour in Extrusion Coating," paper received before the Polyethylene Extrusion Coaters Group, Chicago, June 13, 1963.
9. Conversations with engineers of plastics resin manufacturers.

RECEIVED October 1, 1973.

High Nitrile Copolymers for Food and Beverage Packaging

MORRIS SALAME and EDWARD J. TEMPLE

Monsanto Co., 101 Granby Street, Bloomfield, Conn. 06002

The family of acrylonitrile/styrene copolymers with 60–85% acrylonitrile (AN) content (by weight) exhibits physical and chemical properties suitable for critical food and beverage packaging applications, including carbonated drinks. Tensile strength of 11,500 psi and elongation of 4% assure adequate dimensional stability for containment of the internal pressures of carbonation. Permeation barrier properties of the resins improve as the nitrile content increases, and the high nitrile materials will retain water and CO_2 and will protect the contents against oxygen permeation for expected product shelf life. Tests for dilute solution absorption and for extraction indicate that food flavors will remain essentially unchanged. The results of extensive taste/odor evaluations of several beverages in high nitrile polymer containers confirm the applicability of these resins for packaging uses.

After many years of research, testing, and evaluation, melt processable polymers containing a high degree of nitrile functionality have been developed. These materials possess the excellent barrier, taste, and physical properties required to package foods, carbonated beverages, and other sensitive products (*1, 2*). Because of these properties (*3, 4*) Lopac containers have undergone extensive and successful field evaluations as soft drink containers. (Lopac is a trademark of Monsanto Co.) Sohio has also conducted tests on soft drink bottles prepared from their Barex 210 resin (*5–10*). In addition, DuPont, ICI, and at this writing at least 10 other companies are developing high nitrile polymers for packaging.

This paper describes the physical, chemical, and barrier properties of a new family of high molecular weight copolymers of acrylonitrile and styrene with acrylonitrile functionality in the range of 60–85 weight percent. Data are presented on absorption and extraction along with the results of extensive taste and odor work. The basic structure of these copolymers is shown in Figure 1.

The Nature of Acrylonitrile–Styrene (AN/S) Copolymers

Random copolymers of acrylonitrile and styrene containing less than 30% AN have been well known (11), and many varieties have been sold commercially. The generic material known as SAN, which is a copolymer of 25% AN and 75% styrene, has been sold for many years but has not been used in food or beverage packaging because of its relatively poor barrier and organoleptic properties. There was little or no interest in

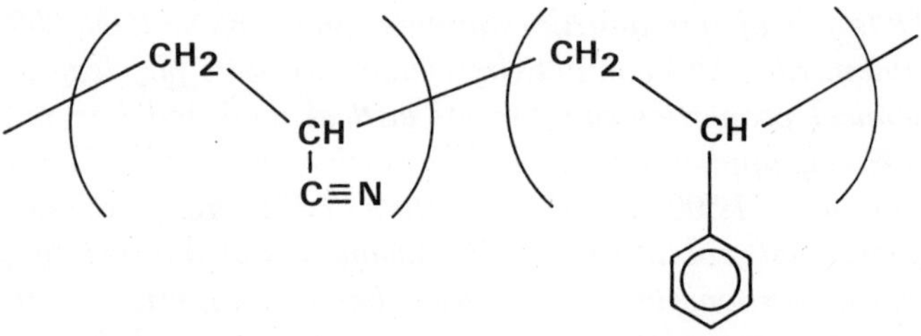

ACRYLONITRILE (AN)   STYRENE (S)

60-85 WT. %   15-40 WT. %

1100-1600 AN UNITS   150-400 S UNITS

Figure 1.   Copolymers of acrylonitrile and styrene used for Lopac containers

copolymers of greater than 30% AN for packaging purposes until the mid or late 1960's. This probably resulted from the difficulty of melt processing polymers with high AN content into useful forms at significant rates and the failure to recognize the excellent oxygen and water vapor properties of these polymers. The recent discovery of the gas and moisture barrier properties of these higher percentage AN materials, coupled with new and improved molding techniques, have made these systems important as packaging materials.

The backbone of acrylonitrile–styrene copolymers containing more than 60% AN is characterized by:

1. High degree of chain-to-chain attraction as a result of polarity (high cohesive energy density) results in chain stiffness and immobility
2. High glass transition temperature
3. Chain order and tight packing (orientation)
4. Chemical inertness.
5. Unwillingness to flow, either in the solid state (cold flow) or molten

6. High molecular weight (*ca.* 100,000)

Most of these properties can be attributed to the high nitrile content of these polymers.

When these AN/S copolymers, are manufactured into Lopac containers, they remain essentially unmodified and in their pure state. Many of the desirable barrier and organoleptic properties of the containers can be attributed to the fact that no plasticizers, rubber, or other common modifiers are incorporated.

Physical Properties of AN/S Copolymers

In many ways the solid state properties of AN/S copolymers of 60–85% AN are typical of high Tg (glass transition), rigid, glassy, amorphous polymers of the polystyrene or SAN class. Unlike polystyrene, however, they have significantly higher tensile strength and are among the highest thermoplastic materials in tensile modulus. Table I lists some physical properties of one of these materials, the 70/30 AN/S copolymer. High tensile strength, high Tg, and high modulus enable the material to contain carbonated beverages under pressure without high creep, distortion, or burst failure.

Container and Material Criteria

The materials for the Lopac container system were developed with the intent of producing packages which would not affect the taste/odor

Table I.   Physical Properties of 70/30 AN/S Copolymer

Property	Value	Advantages in Food/Beverage Packaging
Density	1.13 gm/cc	Light weight.
Tensile strength	11,500 psi	Ability to package soft drink without bulging, bursting (high burst pressure)
Tensile elongation	4.0%	Low creep (carbonated beverage use)
Tensile modulus (stiffness)	650,000 psi	Low creep, low bulging, allows thinner walls (economy), can withstand top load of filling, stacking
Glass transition temperature	108°C	Allows hot fill (juices, drinks), allows *in situ* pasteurization, prevents bulging under adverse storage conditions, etc.
Clarity (light transmission)	90%	Allows product visibility (consumer advantage)
Molecular weight	100,000 MW.	Inertness, toughness

characteristics of products they contained. Criteria were laid out for the several factors which could influence organoleptic values. Figure 2 depicts schematically the factors considered for a typical carbonated beverage. Most of these factors are applicable to many other foods or beverages packaged in a plastic container.

To assign numbers for these variables requires evaluation not only of expected shelf life and exposure conditions but of the susceptibility of specific products to changes affecting taste and odor. Table II lists general criteria established for some of these factors for carbonated soft drinks and malt beverages (12).

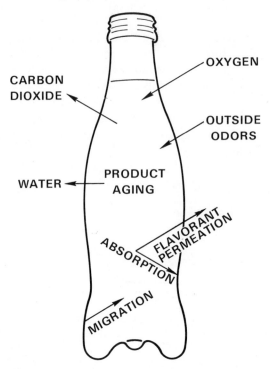

Figure 2. Factors influencing taste/odor of carbonated beverages in polymeric containers

A carbonation loss of greater than 15% would be detectable by taste panel evaluations although it may not be noticeable to an individual not making a definite comparison. Oxygen pickup is a very critical item with beer since a minute amount will result in oxidation of ingredients which causes an objectionable taste. Soft drinks are much less sensitive but still require a good oxygen barrier. The limits for water and alcohol loss have been based primarily on maintaining labeled contents and reasonably consistent fill line levels. The criteria for absorption and

Table II.   Criteria Affecting Taste/Odor of Soft Drinks and Malt Beverages in Polymeric Containers

Characteristic	Maximum Acceptable Change in Commercial Usage
Carbonation	15% CO_2 loss
Oxygen intake	2 ppm—malt beverages
	20 ppm—citrus flavored soft drinks
	40 ppm—cola drinks
Water loss	1%
Alcohol loss	1% of alcohol content
Absorption, permeation	5% of any one flavor peak
Migration	limited by safety and tast/odor considerations

permeation losses have to be rather general because of the great variety of flavorants and wide differences in sensory effects. The term migration covers those influences from package ingredients or trace residuals in the packaging material which may migrate into the food or beverage during the product shelf life. The limits for any extracted chemical are determined by its specific effect on the beverage, by its toxicity, and by government regulations concerning its presence as an indirect food additive.

To apply these package criteria to polymer properties, a conversion was made based on a 10–12 oz. container with a surface area-to-volume ratio of 4.0 (in.²/oz.) and an average wall thickness of 0.030 in. The oxygen, carbon dioxide, and water permeability rates needed to meet these high barrier criteria over a six month shelf life are shown in Table III. Larger container sizes—16, 32, 48 oz. etc.—would permit slightly higher permeability factors for the same bottle criteria, because of their lower ratio of surface-to-volume.

Barrier Properties of AN/S Copolymers

The range of both gas and vapor permeation rates of polymers is well known (13). Table IV lists some typical commercial polymers, along

Table III.   Required Permeation-factors of Polymer to Meet Package Criteria

Permeant	Maximum P-Factor[a]
CO_2	25
O_2	0.8 (malt beverage)
	8.2 (citrus beverage)
	16.5 (cola)
H_2O	2.5

[a] Gas permeation rate—cc/24 hr./100 in²/0.001 in./atm.
H_2O permeation rate—grams/24 hr./100 in²/0.001 in.

Table IV.   O₂, CO₂, and H₂O Permeability Constants of Typical Commercial Polymers at 23°C

Polymer	P-Factor (cc-mil/100 in²-day-atm)		P-Factor (gm-mil/100 in²-day) H_2O (100% RH) (ΔRH 50)
	O_2	CO_2	
Polyvinyl alcohol	~0.002[a]	~0.006[a]	~1200.
Saran (homopolymer)	0.20	0.35	~0.01
Nylon	1.03[a]	4.00[a]	4.7
Kel-F	3.01	12.2	0.05
Polyester (PET)	7.00	30.5	0.77
PVC	8.02	20.5	0.60
SAN (25% AN)	66.8	217.	4.15
Polypropylene	180.	350.	0.10
Polystyrene	416.	1,250.	3.20
Polyethylene	501.	1,500.	0.20
Silicone rubber	~50,000.[a]	~100,000.[a]	19.4

[a] Measured at 0% RH. At high RH, values increase.

with their oxygen, CO_2, and water permeability rates. In general, polymers which are excellent gas barriers are poor water barriers and vice versa. There is also a well defined ratio of CO_2 to O_2 permeation rates of between 2/1 and 4/1. Generally, a material meeting the required O_2 barrier will possess sufficient CO_2 barrier for carbonated beverages.

Using data similar to that in Table IV, we realized that to develop a truly good barrier, one must pay particular attention to the type of substitution on the backbone. Table V shows the great effect upon permeation of various backbone substitutions. The excellent ability of the nitrile group to reduce permeation is seen.

Table V.   Effect of Functional Groups on Permeability

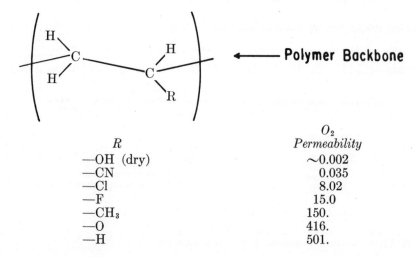

R	O_2 Permeability
—OH (dry)	~0.002
—CN	0.035
—Cl	8.02
—F	15.0
—CH₃	150.
—O	416.
—H	501.

Gas Permeation of Nitrile Copolymers. Table VI gives the O_2 and CO_2 permeation rates of the high barrier AN/S copolymers and for commercial SAN. For further comparison, polystyrene is also listed. O_2 and CO_2 permeation rates are plotted against nitrile content in Figure 3 to show the great influence on permeability of the nitrile group. This barrier is attributed to the extremely high cohesive energy density and polar attractions between chains of the material. These properties create a tight network of chains which is unyielding and unable to readily open for the diffusion step needed by the gas molecule (*14*). Without this step, diffusion is impossible. The AN groups on the polymer backbone, in this sense, can be likened to co-attracting magnets. In some polymer systems, such as poly(vinyl alcohol), polyamides, and cellulose, similar

Table VI.   Gas Permeability of AN/S Copolymers (P-Factor at 23°C)

Polymer	wt. % AN	cc-mil/100 O_2	in²-day-atm CO_2
Polystyrene	0	416.	1250.
SAN	25	66.8	217.
AN/S	60	4.5	7.5
AN/S	67	2.3	5.3
AN/S	82	0.25	0.83
Polyacrylonitrile	100	0.035	0.15

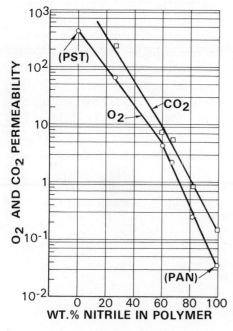

Figure 3.   Gas permeability of AN/S copolymers vs. nitrile content

Table VII.   The Effect of Moisture on the Gas Permeation of Various Polymers

	O_2 P-Factor at 23°C (cc-mil/100 in²-day-atm)	
Polymer	Dry (0% RH)	Film in Contact with water (100% RH)[a]
Poly (vinyl alcohol)[b]	0.002	25.0
Regenerated cellulose (cellophane)	0.02	200.
Poly (caprolactam) (nylon 6)	1.03	6.5
Poly (vinyl acetate)	55.1	150.
SAN (25% AN)	66.8	66.5
AN/S (60% AN)	4.5	4.6
AN/S (82% AN)	0.25	0.20

[a] 50% RH atmosphere downstream.
[b] Lightly crosslinked to prevent film from dissolving in water.

highly polar attractions also exist but are readily destroyed by moisture. Table VII gives the O_2 permeation rate of various polymers in the dry and moist states. In the dry state these other systems are good gas barriers, but with the introduction of moisture, the high chain-to-chain bonding is broken, and gases diffuse readily. Unlike those systems the high AN barrier functionality of the AN/S copolymer is completely unaffected by moisture, even at 100% relative humidity.

Water Permeation of Nitrile Copolymers. While the AN content bears a direct relationship to the gas barrier, the water permeability presents quite an anomaly. If the water permeation of the commercial SAN films (25% AN) is measured, the rate is higher than that of polystyrene. Thus it appears that films with greater AN content have even higher water permeation rates. It was discovered, however, (1, 7) that as the AN content increases there is a shift in permeation, and the higher AN/S materials show water barriers of excellent quality. Table VIII

Table VIII.   Water and Alcohol Permeability of AN/S Copolymers

Polymer	H_2O at 23°C (gm-mil/100 in²-day) P-Factor[a]	50% Solution Ethanol/Water Permeation Rate of Ethanol
Polystyrene	3.20	0.70
SAN (25/75 AN/S)	4.15	0.25
AN/S (60/40)	2.75	0.1
AN/S (67/33)	1.69	0.1
AN/S (82/18)	0.80	0.1

[a] Direct water contact, 50% RH external.

gives the water and alcohol permeation rates, and the water values are plotted in Figure 4.

There is a logical explanation for the increase and then the inversion and rapid decrease of water permeation *vs.* nitrile content. As a small amount of AN groups are introduced on the backbone (*i.e.,* SAN compared with polystyrene), the hydrophilic nature of the polymer is slightly increased (since AN is polar). This causes an increase in water solubility in the film which outweighs any decrease in diffusion caused by the few AN groups. Thus, permeation increases. As more AN groups are introduced water solubility reaches an equilibrium value, but water diffusion begins to rapidly slow down because of the chain packing and resistance of the chain to open into diffusion paths as explained previously. Thus, permeation decreases. This inversion is not seen, however, for alcohol or other organic liquids where permeation is directly proportional to the AN content as in the case of gases.

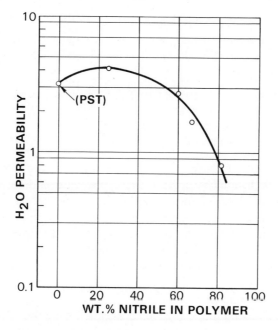

Figure 4. Water permeability of AN/S copolymers vs. nitrile content

Absorption and Permeation

Experiments on absorption and permeation of flavorants have been carried out in high nitrile barrier containers with a number of organic compounds which represent a variety of chemical functionalities. Table

IX lists the results of some of these experiments, comparing the nitrile materials with polyethylene, the most widely used plastic container material, and poly(vinyl chloride), which is being used for a number of food packaging applications. Note that in all instances there is an order of magnitude difference between the Lopac container and the other two. For flavorants, which are usually present in very low concentrations, this dilute solution test is probably more significant than a standard permeability test which only measures weight losses of the pure ingredient.

Table IX.   Dilute Solution Absorption Values for Polyethylene, Poly(vinyl Chloride), and Nitrile Polymer

		Percent Loss of Organic After 1 Month at 120°F		
Test Compound (0.1% Concentration)	*Chemical Functionality*	*Poly-ethylene*	*PVC*	*Lopac*
Menthol	alcohol	30	25	<1
Citral	aldehyde	53	30	4
Methyl salicylate	ester	69	20	<1
Carvone	ketone	80	40	<1
Menthone	ketone	95	60	<1
Dipentene	hydrocarbon	98	15	<1
Chloroform	halogenated H-C	99	20	5
Anethole	ether	100	75	<1

Polymer Migration Influences

The migration of package ingredients directly into a food product is often difficult to analyze instrumentally because of interference from food ingredients. Some of these analyses have been made, but it is generally preferable to use food simulating solvents listed in FDA regulations and to carry out extraction tests under the conditions described.

Tests have been conducted with Monsanto high barrier nitrile resins using the common food simulating solvents (Table X) plus some typical beverages. Conditioning times and temperatures were based on applicable FDA regulations and guidelines (16).

Analytical procedures sensitive to 2 ppm for styrene and 0.05 ppm or less for other items were used for examining the extracts. Even under these exaggerated exposure conditions no detectable levels of the monomers, of the polymer, or of other potential residuals were observed. The materials are truly "non-food-additive" by the FDA definitions. Hydrogen cyanide was included in the list of substances for analysis since it can be present at low levels in commercial acrylonitrile monomer, and it has been reported as a thermal decomposition product of acrylonitrile polymers. As shown here, it is not detectable in extracts by tests sensitive to

Table X.   Extraction of 10 oz. Lopac Containers

Extracting Solvents

Acetic acid (3%)	Ethanol (8% in H_2O)
Water (distilled)	Ethanol (25% in H_2O)
Heptane	Ethanol (50% in H_2O)

Results of Analyses

Examine for	Test and Sensitivity	Results
Acrylonitrile	G.C. at 0.05 ppm	Not detected
Styrene	G.C. at 2 ppm	Not detected
Hydrogen cyanide	Fluor at 0.02 ppm	Not detected
Oligomers	As total non-volatiles N_2 analysis at 0.05 ppm [a]	Not detected
Process aids (including catalyst)	Various tests at 0.05 ppm	Not detected

[a] All nitrogen in non-volatile extracts is measured by C-H-N analyzer and calculated back as oligomer.

20 ppb. This procedure responds to CN⁻, cyanogen, and other cyano-genetic substances and thus demonstrates the low potential for extraction of the cyanide moiety. We believe that the absence of monomer extraction evidenced here resulted from (1) the purity of the polymers used and (2) the very high diffusion barrier of the polymer matrix.

Sensory Tests

These objective, quantitative tests have shown that nitrile containers should protect the taste and odor of packaged foods and beverages. But the primary consideration in judging overall package performance, once safety is assured, rests on the subjective evaluations of taste, odor, and appearance. It is well known that the animal senses, in many instances, are far more sensitive than the best instruments and also are capable of integrating the individual effects of the several influences on product quality.

Among the most sensitive taste-odor test methods is the Triangle Difference Test, a Forced Choice test, generally recommended when there are only slight differences (*17*). In our work a normal taste panel consists of twenty judgements, although occasionally a smaller panel of especially sensitive individuals may be used for specific problems. A determination of a significant difference between the control and the test sample was based on a 95% confidence level—in this case 11 or more correct judgements out of 20. The comments of the panelists were also analyzed for any additional information.

Triangle tests were carried out under a variety of conditions. Let us first consider testing beverages to which trace amounts of the two monomers—acrylonitrile and styrene—and hydrogen cyanide were added directly. The objective was to determine whether these sensory procedures would detect the presence of such chemicals if they were to extract from the polymer at levels below our analytical sensitivity. The data in Table XI show that styrene monomer is easily detected in cola drinks at levels well below our instrumental abilities. Acrylonitrile, in extracting solvents, can be measured readily by gas chromatography at very low levels, but our sensory tests, where we are dealing with a complex cola drink, are less sensitive and add little to our knowledge. There is an extremely sensitive fluorimetric method for hydrogen cyanide in pure solvents, and our tests indicate that there are a significant number of people who can detect it at comparable levels in cola.

Table XI.   Taste/Odor Detection of Additives in Cola Beverage

Additive	Minimum Detectable in Solvents	Detection by Triangle Test in Cola
Styrene	2 ppm	1.0 ppm—99.9% Confidence
		0.5 ppm—99% Confidence
		0.2 ppm—99% Confidence
Acrylonitrile	0.05 ppm	0.1 ppm—Not detectable
		0.05 ppm—Not detectable
Hydrogen cyanide	0.02 ppm	0.1 ppm—99.9% Confidence[a]
		0.05 ppm—99% Confidence[a]
		0.03 ppm—95% Confidence[a]
		0.02 ppm—No significant difference[a]
		0.01 ppm—Not detectable[a]

[a] Select panel.

Similar Triangle tests have been conducted with a variety of beverages packed in Lopac containers made during experimental runs. The beverage was stored in the test container and in glass controls at 100°F for seven days. The same procedure was used for long term tests at room temperature for 1, 2, and 3 month intervals. Table XII presents the results of these tests. The data show the general compatibility of this container system with both carbonated and noncarbonated products. With most of these carbonated drinks no significant difference was found under a variety of test conditions. Two different brands of orange soda yielded the same results—a preference for the drink packaged in the Lopac container. It appears from other work that the glass surface catalyzes a reaction of a limonene ingredient in orange flavors which reduces the flavor impact.

Table XII.   Taste Test Results

Product and Test Condition	Triangle Test Result vs. Glass Control

Coca Cola
  1 wk. at 100°F
  1, 3, 6, 12 mo. at 73°F
	No significant difference

Dr. Pepper
  3 mo. at 73°F
  6 mo. at 73°F
	No significant difference

7-Up
  1 wk. at 100°F
  1, 2, 3 mo. at 73°F
	No significant difference

Fresca 1 wk. at 73°F	No significant difference

Cotts Ginger Ale
  1 wk. at 100°F
  1, 2, 3 mo. at 73°F
	No significant difference

Cotts Orange Soda 3 mo. at 73°F	Significant difference. Product in Lopac bottle preferred.

Fanta Orange Soda
  6 wk. at 73°F
  3 mo. at 73°F
	No significant difference
	Significant difference. Product in Lopac bottle preferred.

Hi-C Orange, Grape
  1 mo. at 73F
  1 mo. at 100°F
	No significant difference

Coors Beer 3 wks. at 40°F	No significant difference

These results, considered in relation to the direct addition tests of monomer and hydrogen cyanide in the previous table, demonstrate that there is no reason to expect styrene monomer extraction into soft drinks, even at levels well below those we can measure analytically. They also reinforce our hydrogen cyanide data. Further, they indicate that these beverages are not more extractive of Lopac containers than the normal simulating solvents. The tests confirm the chemical safety of the containers as beverage packages.

Profile Analysis

Another valuable tool in food package development is the taste/odor Profile Test (*18*). Developed by the Arthur D. Little Company, this method uses a small panel of four to eight people specially trained in the procedure. Their analysis develops a full descriptive terminology for the product in question and assigns a numerical rating to each component of the taste/odor complex.

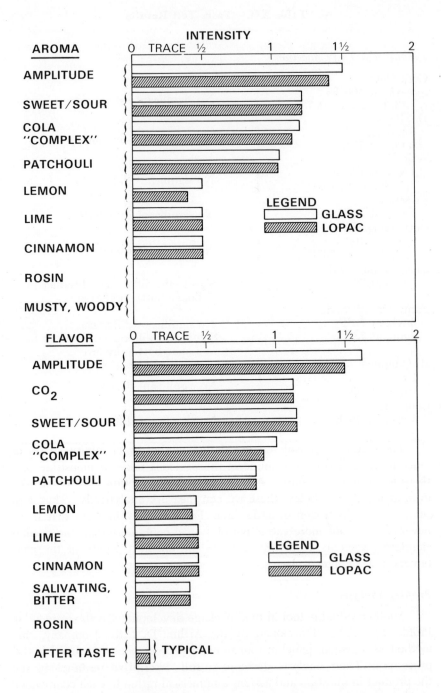

Figure 5.   Profile of cola drink in Lopac and glass containers

Figure 5 illustrates graphically the Profile Analysis of a cola product. The comparison is based on evaluation of the beverage after storage in both Lopac and glass bottles at 90°F for 3 weeks. From this analysis we can generalize that the product, when packaged in Lopac, retains essentially the same taste profile as in the glass controls. If the differences were significant or objectionable additional profile testing could be done to isolate and identify the principal factors. One important inference that can be drawn from this profile and the Triangle tests is the absence of foreign odor or flavor contributions from this container system. Again, this is caused by the non-extractive nature of the polymer and also by its barrier properties which seal the contents from atmospheric oxygen and from stray odors in the test environment.

Conclusion

Our studies of the absorption, permeation, and extraction properties of containers produced from high nitrile barrier resins have demonstrated that they meet or surpass the basic criteria established for retention of taste and odor characteristics of carbonated soft drinks. Sensory tests, which can isolate and identify end results as well as integrate collective effects, have confirmed this judgement and have established the general compatibility of these containers with a variety of beverage products from a taste and odor standpoint. Furthermore, these materials have the excellent physical properties required for containers which will find wide use in food and beverage packaging.

Acknowledgment

The authors thank S. P. Nemphos of the Monsanto Co. for preparing the copolymers used in these studies.

Literature Cited

1. Trementozzi, Q. A., U.S. Patent 3,451,538 (1969).
2. Trementozzi, Q. A., U.S. Patent 3,540,577 (1970).
3. Salame, M., "Transport Properties of Nitrile Polymers," *Amer. Chem. Soc., Div. Org. Coat. Plastics, Preprints* (1972) **32**, (2).
4. Salame, M., Schickler, W. J., Steingiser, S., "New High Barrier Polymers, Their Properties and Applications," Polymers Conference Series **6**, University of Utah (July 20, 1972).
5. Solak, T. A., Duke, J. T., U.S. Patent 3,426,102 (1969).
6. Ball, L. E., Greene, J. L., *Encycl. Polym. Sci. Technol.* (1971) **15**, 321-353.
7. Hughes, Idol, Duke, Wick, *J. Appl. Polym. Sci.* (1969) **13**, 2567-2577.
8. Hughes *et al.*, *Polymer Reprints* (1969) **10**, (1), p. 403-410.
9. Blower, K. E., Standish, N. W., Yanik, R. W., *SPE J.* (Nov. 1971) **27**, p. 32-36.

10. Blower, K. E., Yanik, R. W., *SPE J.* (March 1972) **28**, p. 35-38.
11. Ziemba, G. P., *Encycl. Polym. Sci. Technol.* (1964) **1**, 425-435.
12. Salame, M., "No Customer Complaints On Taste When You Pick The Right Plastic," *Package Eng.* (July 1972) pp. 61-66.
13. Salame, M., *Polymer Preprints* (1967) **8** (1), 137.
14. Crank, J., Park, G. S., "Diffusion in Polymers," Academic, London, 1968.
15. Morgan, P. W., *Ind. Eng. Chem.* (1953) **45**, 2296.
16. FDA Reg. **121.2514, 121.2526.**
17. *Book ASTM Stand.*, **STP 434.**
18. Sjöström, L. B., Cairncross, S. E., Caul, J. F., "Methodology of the Flavor Profile," *Food Technol.* (1957) **XI** (9), zo-25.

RECEIVED October 1, 1973.

Flexible Packaging and Food Product Compatibility

JOHN E. SNOW

Champion Packages Co., Minneapolis, Minn. 55414

Flexible packaging films and laminations are suitable for holding various food products depending on the moisture, gas, grease, and light barrier properties of the component films. There are various packaging levels for ketchup such as cellophane/polyethylene and foil-containing packaging. Polypropylene in chunk cheese packaging eliminates product loss by eliminating flex cracking through the hinge effect of oriented polypropylene. Flex crack resistance of packaging films is determined by a variation of the MIT folding endurance test. Paper, cellophane, polyester film, aluminum foil, poly(vinylidene chloride), poly(vinyl chloride), and polyethylene are also commonly used packaging materials. Specific examples of packaging for mustard, biscuit mixes, and drink mixes are given. Test methods are described to show how and why certain films are used for mechanical, chemical, and barrier properties to avoid over- or under-packaging.

Since flexible packaging materials are by nature relatively thin, light weight, and generally contain minimum amounts of materials per unit weight packaged, each element of the package structure must perform well for the package to be a success. Much testing must be done to prove that the levels of mechanical and chemical protection are adequate for the commercial market before the package can be released with assurance. All film structures must meet FDA requirements, but these factors will not be discussed here.

This paper organizes some of the "how" and "why" of flexible package–food product interactions by discussing a few specific examples of food packaging development, as well as some overall factors of packaging material application.

Compatibility Testing

The method used for testing compatibility and shelf life consists of:

1. Making packages
2. Filling the packages with the food product being tested
3. Holding the filled packages at 100°F in a circulating air oven for designated periods of time
4. Opening the packages and testing for package integrity and changes in food color, flavor, odor, or taste.

The flexible film products used in these experiments were all constructed for heat seal closure. Packets were made by heat sealing three sides of two 2½ inch squares of the subject film together face to face. The heat seal was ¼ inch wide along each edge.

The packet thus formed was filled with the food product to be tested and the open side was heat sealed. Heat seals were usually made at 350°F jaw temperature, 40 psi pressure with 0.5 second dwell time; the exact conditions necessary depend on the materials used and caliper. This produced a packet with an 8 sq. in. total area.

These packets, usually six to twelve for a given film, were placed in the 100°F circulating air oven on edge in an open top box. In this manner both liquid and vapor space tests are made simultaneously in case there are unexpected differences.

Sample pouches were then opened and both the food product and the film were examined at appropriate time intervals. The food product was examined for changes in texture, color, odor, and taste. The packaging material was checked for color changes, delamination, seal strength, etching of foil (if present), and any other changes noted.

When moisture loss was to be determined, each packet was weighed initially and at intervals of 2–4 weeks through the holding time. Moisture loss rates were calculated on the basis of grams/100 sq. in./24 hours (excluding the first 24 hours which constitutes a packaging material conditioning time).

Product moisture gain was checked by the same basic procedure, except that the packages were stored in a humidity cabinet at 100°F/95% RH. Weight gain rate was determined by comparing original weight of packet with the weight after the given time ended and calculating the daily rate in grams/100 sq. in. (Again the first 24 hours was excluded because of the packaging material conditioning factor).

Properties of Flexible Packaging Films

There are four major causes of incompatibility between a packaging material and the contained product:

1. Packaging material permeability
   a. Into the package—oxygen, water vapor, light
   b. Out of the package—water vapor, flavors, oils, etc.

2. Chemical attack of the contents on the packaging material, perforation

3. Materials problems. Components of the packaging material dissolving in the product, changing taste, odor, shape, texture, etc.

4. Mechanical breakage

Among the commonly used flexible packaging materials, aluminum foil probably provides the most complete permeation barrier while paper is the most permeable. Although aluminum foil provides a barrier to moisture, gas, grease, and light, it usually needs protection from the contents of the package and from the environment since it is a soft metal and subject to chemical attack.

Water vapor, oxygen, and grease are three of the most troublesome materials to contain or exclude from a package. Table I indicates the ability of the most commonly used packaging films to resist passage of water, oxygen, and grease.

Moisture permeability (MVTR) of packaging films or film laminations is generally determined either by weighing the water passing

Table I.   Properties of Flexible Packaging Films

	WVTR	Gas Transmission	Grease Resistance	Other Properties
Paper	high	high	0	—
Waxed glassine	low	low	excellent	—
Aluminum foil	0[a]	0[a]	excellent	light barrier
N/C cellophane	high	moderate (wet)	excellent	tears easily
PVDC-coated cellophane	low	low	excellent	tears easily
Cellulose acetate	high	high	medium	—
Polyester	medium/high	medium	excellent	tear resistant
PVDC-coated polyester	low	low	excellent	tear resistant
Oriented polypropylene	low	high	excellent	flex crack resistant
Nylon	high	medium	excellent	formable, crack resistant
Polyethylene	low	high	medium	sealable, low cost
High density polyethylene	low	high	excellent	sealable
Vinyl	medium	medium	excellent	—

[a] Except for possible pinholes in foil of thin gauges.

through a specified area of film, as in ASTM E–96 test method, or by measuring the diffusion rate of water vapor through a film by instrumental means. Wood (1) and Heisler (2) have discussed these methods in some detail. Gas transmissions have generally been determined by the ASTM D–1434 method.

Paper is one of the oldest and most commonly used packaging materials. Generally it is used to keep a product clean and for mechanical strength when combined with other materials. It does not protect a product from atmospheric change, but only from mechanical contaminants such as dirt. Coated papers are much more functional. Waxed papers fall into this category and provide much better protection from moisture and, in some cases, from gas (oxygen) transmission.

Glassine is a highly processed paper product with some grease resistant properties. When laminated with a suitable type of wax, it gives excellent protection from both moisture and gas transmission. For this reason it is used in packaging dry cereals.

Cellophane is an old and respected packaging material which has been improved over the years. The two general types are coated with nitrocellulose (N/C) and polyvinylidene chloride (PVDC), respectively. Nitrocellulose-coated cellophane is "moisture proof" and useful for packaging dry products. It does not exclude oxygen or moisture completely, but for noncritical products it is entirely satisfactory. Baked goods are often packaged in this breathing type film. It is often used for cookies, candies, and rolls because its lack of taste and odor makes it very compatible with these products.

N/C-coated cellophane allows moisture loss or gain by a packaged product. When cellophane is PVDC coated, however, both the gas and moisture transmission rates are reduced, which lengthens shelf life. Products such as nuts, potato chips, and the like are much better packaged in PVDC-coated cellophane because of lower gas and moisture transmission rates. Light, especially UV, can hasten rancidity development in oil or fat-containing products of this type, so the partial UV absorbing characteristics of PVDC help avoid product change in yet another way and make the package more compatible with the product.

Although polyethylene film is much used and gives a high degree of moisture protection, it provides almost no oxygen (gas) protection. It is relatively odorless, tasteless, and generally inert so it is a useful component which is compatible with many products. It has been used for bakery product wrapping with excellent results. However, its fogging tendency is objectionable in some applications where cellophane would be more useful. Bread, once packaged in waxed paper and then in cellophane, has been converted almost completely to printed polyethylene bags in the last several years because of the increased moisture barrier, toughness, and freedom from tearing. The bag is also reusable.

Oriented polypropylene is a particularly compatible film for chunk cheese packaging. A satisfactory film can be flexed up to 5–10,000 times without failure compared with less than 1,000 for polyester film. A nearly complete changeover has been made to this type film for chunk cheese packaging during the past ten years. With PVDC-coated polyester/polyethylene composite film, only short shipping distances could be tolerated before 10–30% of the packages opened, causing the possibility of mold development. With either properly designed, Saran-coated, oriented polypropylene/polyethylene or with oriented polypropylene/polyethylene/K cellophane/polyethylene (Curpolene 200) the leaker rate is 0–5% even during long distance shipping. In this case, compatibility was achieved by eliminating permeation caused by film failure (*see* Table II).

While oriented polypropylene has a hinge effect which resists cracking and has good moisture barrier properties, it does not have gas barrier properties. This problem is solved by using a PVDC coating. During a detailed study of several films for cheese packaging, a test method was developed for checking the flex crack resistance of coated polypropylene and other films.

Table II.   Flex Testing of Cheese Packaging Films

Packaging Films	Flex Cycles to Crack
110ga oriented polypropylene Saran coated/2 mils PE	5–20,000
50ga oriented polypropylene/PE/PVDC-coated cellophane/PE	5–20,000
50ga polyester, Saran-coated/2 mil PE	1,000

MIT Flex Endurance Tester

Although the MIT Flex Endurance Tester (3) is generally used for paperboard and papers, the test procedure was changed so it would provide useful data on flex crack properties. In this test a piece of the film 1 inch wide is folded to ½ inch width with the inside of the film folded in (Figure 1). It is then flexed in the normal fashion on the MIT Tester for the desired number of cycles. The sample is removed and checked for a crack at the flex–fold point by applying a drop of red-dyed turpentine oil and checking for penetration with white blotting paper. While polyester/polyethylene film samples will show no failure after flat film flexing, the folded sample will crack in less than 1,000 cycles. A satisfactory PVDC-coated, oriented polypropylene/polyethylene film will not show penetration before approximately 5,000 cycles.

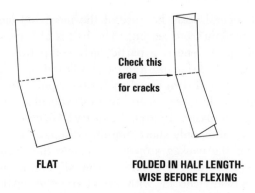

Figure 1.   Flex crack test for film using the
MIT Folding Endurance Tester

Specific Package/Product Compatibility Problems

The different food products discussed below illustrate compatibility/ incompatibility problems and solutions in flexible packaging.

Ketchup. The ketchup, tomato sauce, barbeque sauce type products are difficult to package. The compatibility factor is guided by consumer requirements and the expected shelf life or market pattern.

Ketchup has been packaged successfully for the short time span market in paper/polyethylene or N/C-coated cellophane/polyethylene. However, moisture is gradually lost from the package and oxygen seeps into the package. Thus, the ketchup dries out and the entire contents of the package turn a dark, black plug ketchup color because of the reaction with oxygen. This type of product may satisfy the nondemanding consumer if a shelf life of only 3 months is required, but it is not really a good compatible package. The use of PVDC-coated cellophane/ polyethylene extends the shelf life by perhaps another 2 months because of the added barrier against oxygen and moisture.

A far more compatible system was a package film containing aluminum foil. In this case, the aluminum foil was protected from the acid

Table III.   Ketchup Packaging Films

Film	Storage Time (months)	Product Condition
35# Paper/2 mil PE	3	dark color, dried out
PVDC-coated cellophane/2 mil PE	5	darkening, dried, but still usable
N/C-coated cellophane/2/3 mil PE/35ga foil primer and/or adhesive/1½ mil MDPE	12	bright color, no flavor or weight loss

components of the product by a primer and/or adhesive over the foil and under the polyethylene sealant layer. A film-containing, nitrocellulose-coated cellophane/polyethylene/aluminum foil/primer and/or adhesive/MD polyethylene film has a shelf life of 9–18 months with virtually no moisture loss and complete color and flavor retention. Without the protective primer or adhesive coating on the aluminum foil, however, the acid components would penetrate the polyethylene layer, react with the aluminum to produce hydrogen gas, puff up, and destroy the pouch while completely changing the color, texture, and taste of the product. Package tests, following the method described above, have shown this in repeated cases. The specific structures in Table III summarize the overall picture.

Mustard. Mustard has been packaged in PVDC-coated cellophane/polyethylene pouch material which was recognized to have a 3–6 month shelf life. In this case, the moisture loss rate was diminished by using the PVDC-coated cellophane. The main drawback of this package was the moisture loss, since the flavor was not markedly changed by the oxygen traces allowed by the PVDC cellophane coating.

Table IV.   Mustard Packaging Films

Film	*Storage Time*	*Product Condition*
195 PVDC-coated cellophane/2 mils white PE	4 months	some drying, usable
Ketchup packaging film—(N/C-coated cellophane/2/3 mil PE/ 35ga foil primer and/or adhesive/ 1½ mil MDPE)	2–4 weeks	package destroyed
N/C-coated cellophane/2/3 mil PE/ 35ga foil/adh/50ga polyester/PE	9 months	as packaged

Further experiments to lengthen the storage life of mustard packaging showed that it was a much more difficult problem than ketchup. An aluminum foil package which contained ketchup for long periods, was etched through in 2–4 weeks by mustard. A better barrier was needed to keep the acids and essential oils from penetrating the polyethylene and attacking the aluminum foil.

The structure—N/C-coated cellophane/polyethylene/aluminum foil/adhesive/50ga polyester/adhesive/polyethylene—had a shelf life of over 6–9 months with no moisture, flavor, or color loss. This barrier system contained the product successfully (*see* Table IV).

Red Meat. The in-store wrapping of fresh meats is another packaging system with interesting compatibility requirements. If red meats are packaged in a heavy polyethylene film, 2 mils thick for example, the meat will turn dark. If, however, a ¾ mil vinyl film with high oxygen transmission rate is used, the oxygen will keep the "bloom" in the meat, retaining the red color and appearance. Here, the high gas transmission is needed for marketability. Cellophane also transmits oxygen and can be used to package red meat, but its tear resistance is low. It has therefore been replaced by the vinyl type film which forms a more compatible system.

Drink Mixes. Drink mixes are often packaged in highly decorative printed paper/poly/aluminum foil/polyethylene packaging films. If well protected, the flowing granular mix is retained without caking for many months. If these packages are shipped on edge, however, the flexing may produce leakers, and the drink mix may cake because of its hygroscopic nature. One solution to this problem has been to laminate an 0.5 mil oriented polypropylene film into the structure so that there are no complete flex cracks through the structure. This takes advantage again of the hinge effect of oriented polypropylene. The improved package/product compatibility results in fewer complaints and decreased product losses.

Biscuit Mixes. Biscuit mixes are often packaged in printed paper/heat sealant-coated pouches. At one time the paper was coated with polyethylene and PVDC (as the sealing medium) for grease resistance, since many biscuit mixes contain up to 12% shortening. Many grease spots appeared on pouches because of cracks appearing in the brittle PVDC

Table V.   Biscuit Mix Packages

Packaging Material	*Product Condition*
40 # Paper/10 # PE/6 # PVDC	packages greased through and broke during shipment
40 # Paper/15 # ionomer resin	packages clean with no breakage in shipment

coating during shipment. In addition, the seals had to be made at a high temperature and slow speed because of the PVDC melt temperature.

Grease resistant ionomer resins and structures made with coated paper/polyethylene/ionomer and paper/ionomer were then considered. Holding and shipping tests showed no leakers or greasing through with these packages. The greater flexibility and crack resistance of the ionomer provided the needed compatibility. In addition, it was possible to obtain faster packaging machine speeds as well, because of the lower sealing temperature required (*see* Table V).

Trace Solvent Removal. Several papers were written by Nadeau (4) and by Gilbert (5) and co-workers on gas chromatographic methods for determining solvent traces remaining in flexible packaging films after printing or adhesive lamination. With proper equipment and techniques, residual solvent can be brought to less than 500 ppm, where there is generally no detectable transfer to the product. If insufficient drying or inappropriate temperatures, air flow rates, or machine speeds are used, larger quanties of solvent can be left in the film. If this happens, it can transfer from the film to the food product causing a change in flavor and/or aroma.

For this reason, many converters use the gas chromatograph as a quality control instrument for checking retained solvent in each run of film produced. Even more interesting are the systems of water-based inks now being tested which will eliminate this particular type of compatibility/contamination problem where they are practical. The solventless UV cure inks discussed by Carlick (6) and recently introduced for offset printing also provide a method for largely eliminating the possibility of this type of problem. These last two methods also will help to reduce some forms of air pollution when the ink developments are completed and the materials are ready for use in rotogravure and flexographic printing.

Mechanical Breakage. Mechanical breakage of packages can occur for several reasons. The materials may not have sufficient strength (heavy enough paper or film), puncture resistance may be insufficient, or one part of the structure may be too weak (or too strong) for the particular application.

When a small paper/polyethylene flour pouch broke as it dropped off the conveyor line before being packed into a prepared menu type of product, a different factor was found. Investigation showed that the heat seals were good, and the breakage was occurring along the edge of the seal. Tests were then made by dropping filled bags made with paper/polyethylene with variations in bond strength between the paper and polyethlyene coating. The results of the drop tests showed that a loose bond between the polyethylene and the paper eliminated the breakage problem by allowing a slight give upon impact. The use of paper treatment to produce a tight bond had been wrong for this particular application.

From these examples, it can be seen there are a great many factors to be considered in producing a satisfactory degree of package/product compatibility for a given product and consumer group.

Acknowledgment

The author thanks Robert G. Kessler, David J. Sachi, Dee McKee, and the laboratory staff for their technical input and aid in these investigations.

Literature Cited

1. Wood, R. C., *Mod. Convert.* (1972) **16**, (4), 32–36.
2. Heisler, J., *Gravure* (1971) **17**, (8), 6–10, 32, 34, 38.
3. *Book ASTM Stand.*, "Folding Endurance of Paper by the MIT Tester," **ASTM-D-2176.**
4. Nadeau, H. G., Neumann, E., *Mod. Packag.* (1964) **37**, (2), 128.
5. Gilbert, S. G., Oetzel, L. I., Asp, W., Brazier, I. K., *Mod. Packag.* (1965) **38**, (5), 167.
6. Carlick, D. J., *Mod. Packag.* (1972) **45**, (12), 64.

RECEIVED October 1, 1973.

Irradiation of Multilayered Materials for Packaging Thermoprocessed Foods

JOHN J. KILLORAN

U.S. Army Natick Laboratories, Natick, Mass. 01760

The mechanical, chemical, and thermal properties of multilayered flexible materials were improved for packaging thermoprocessed foods. Four multilayered materials were irradiated (gamma) at 1–18 Mrad and exhibited mechanical properties superior to the unirradiated materials. The seal strength of a typical, adhesive-bonded material of poly-(ethylene terephthalate), aluminum foil, and the ethylene–butene copolymer increased by 43% after irradiation at 8 Mrad. Bond strength between the ethylene–butene copolymer and the aluminum foil increased by 600%. Chemical and physical analyses of test specimens indicated that the strong adhesion among layers was not caused by the mechanical interlocking of layers but by the formation of primary chemical bonds extending across the interface. The improvements in the multilayered materials caused by the irradiation were also seen in irradiated pouches filled with beef, vacuum sealed, and retorted.

Foods packaged in flexible containers and processed for commercial sterilization are a part of the trend toward improved quality in convenience foods. The retorted pouch has the package integrity of the metal can coupled with the food quality, reheating, and serving convenience of frozen, boil-in-the-bag foods (*1*). Its use for combat rations has especially interested military R&D personnel because of the functional advantages over the metal can (*2*).

This paper describes an irradiation curing method which improves the mechanical, chemical, and thermal properties of multilayered flexible materials, increases the bond strength among the adhesively bonded layers, and provides flexible packages that can withstand the thermoprocessing stresses and subsequently protect their contents during normal handling and storage.

Experimental

Materials. The single films and multilayered materials examined are listed in Tables I and II, respectively. The adhesive (10μ thick) between any two layers was a two-component curing, epoxy–polyester adhesive, Adcote 503A, Morton Chemical Co. Pouches (11.5×17.8 cm) were made from the multilayered materials by heat sealing the sides and bottom under the optimum heat sealing conditions (3).

Table I.   Commercial Films for Thermoprocessing

Polyethylene, 0.960 g/cc
Ethylene-butene copolymer, 0.950 g/cc
Blend of ethylene-butene copolymer and polyisobutylene
Polypropylene, 0.905 g/cc
Blend of polypropylene and propylene ethylene copolymer
Poly(ethylene terephthalate)
Polyiminocaproyl (nylon 6)
Blend of polypropylene and ethylene vinyl acetate copolymer

Table II.   Multilayered Flexible Materials

Pouch Number	Material	Thickness (cm × 10³)
1	Poly(ethylene terephthalate)	1.3
	Aluminum foil	0.9
	Ethylene–butene copolymer	8.0
2	Poly(ethylene terephthalate)	1.3
	Aluminum foil	0.9
	Ethylene–butene copolymer—polyiso-butylene blend	8.0
3	Polyiminocaproyl (nylon 6)	2.5
	Aluminum foil	0.9
	Polyiminocaproyl (nylon 6)	2.5
	Ethylene–butene copolymer	8.0
4	Poly(ethylene terephthalate)	1.3
	Ethylene–butene copolymer	5.0
5	Polyiminocaproyl (nylon 6)	2.5
	Aluminum foil	0.9
	Polypropylene—propylene ethylene copolymer blend	8.0
6	Poly(ethylene terephthalate)	1.3
	Aluminum foil	0.9
	Polypropylene—ethylene vinyl acetate copolymer blend	8.0

Irradiation Conditions. The gamma (cobalt-60) radiation facility and the source calibration are described by Holm and Jarrett (4). Irradiation temperature was 21 (initial) – 40°C (final). The gamma source was calibrated with the ferrous sulfate/cupric sulfate dosimeter for a dose rate of 8×10^2 rads per second. Pouches were fabricated from multilayered materials and then irradiated while empty. The container used to hold the multilayered materials and the empty pouches during irradiation was a large size, flexible package that was sealed under vacuum prior to the irradiation.

Testing Methods. Melt index, seal strength, bond strength, and Vicat softening point tests were performed according to standard ASTM methods. The irradiation-induced chemical changes in the multilayered materials were determined by ATR infrared spectroscopy using a Beckman IR10 grating spectrophotometer. Melting temperatures of the polymeric films used as components of the multilayered materials were determined by differential scanning calorimetry.

Thermal Sterilization of Pouches of Beef. Methodology for food thermoprocessing in cylindrical metal containers assured food sterility in flexible packages. Beef slices (1.25 cm thick) were steam cooked to an internal temperature of 72°C and vacuum sealed in pouches to give a fill of 120 g. These pouches of beef were processed in a standard retort with complete water circulation and a superimposed air pressure of 1.7×10^5 Pa. The retort schedule—a 40-minute cook at 118°C plus come-up time followed by a 30-minute cooling time—achieved a F_o (lethality value) of 6.

Results and Discussion

Background. No single flexible material has all the chemical, physical, and protective characteristics necessary to meet all the requirements for a container for thermoprocessed foods. Therefore, flexible packages have been fabricated from multilayered materials. These packages must be easily heat sealed and able to: withstand the thermoprocessing temperatures without melting, delaminating, or losing seal strength; withstand handling and shipment hazards; protect the contents from microbial or other contamination; provide an oxygen and moisture barrier; and be inert to the package contents.

The commercial polymeric films (Table I) that are used as the outside layer of multilayered materials for thermoprocessed food packaging are poly(ethylene terephthalate), polyiminocaproyl, or polypropylene. The other five films listed in Table I or polypropylene are used for the food-contacting layer.

Thermal analysis showed that polyethylene, ethylene–butene co-polymer, and the blend of polyethylene and polyisobutylene should be classified as borderline when used as components of the pouches that are exposed to 118°C for 40 minutes during the thermal processing of foods. Polyethylene with a density of 0.960 g/cc melts from 110–141°C, the peak temperature being 134°C. Both the ethylene–butene copolymer and the blend of polyethylene and polyisobutylene melt from 108–128°C, the peak temperature being at 125°C. Also, orientation stress analysis has shown that these crystalline polyolefins, when exposed to thermal processing and/or heat seal temperatures, tend to shrink and pucker in the restrained state of the multilayered material. This leads to delamination, especially in large-size pouches. Polypropylene (melting point, 168°C) is satisfactory as a component of a multilayered material for packaging thermoprocessed foods. However, it has one disadvantage for storing these foods in that it has a relatively high second-order transition point. Only a slight reduction of this temperature is achieved by biaxial orientation and heat setting of polypropylene or by blending the polypropylene with a propylene–ethylene block copolymer.

Multilayered materials owe their properties and behavior to the properties of and the interactions between the components (5). Each of the two or more components contributes its particular property to the total performance of the multilayered material. For example, in Pouch 1, Table II, the aluminum foil provides high oxygen and water vapor permeability resistance, poly(ethylene terephthalate) provides structural strength and stiffness, and the ethylene–butene copolymer provides a heat sealable layer. If the components of the multilayered materials interact then the whole would be something different than the sum of its parts. In other words, the properties of the components of the multilayered materials are not independent of one another but rather are interdependent.

In a well-bonded material consisting of poly(ethylene terephthalate), adhesive, aluminum foil, adhesive, and poly(ethylene terephthalate), the aluminum foil contributes significantly to the load-bearing ability. It can undergo ductile plastic stretching, attaining a high degree of elongation without delaminating or tearing from the poly(ethylene terephthalate). In a poorly bonded material, the aluminum foil will tear and delaminate from the poly(ethylene terephthalate). In this case the role played by the adhesive in regard to its tensile load-carrying capacity is important (6).

Each of the multilayered materials of Table II, in pouch form, met the retortability requirements. Examination of the pouches after this test showed that no delamination occurred among the layers. However, microscopic examination of specimens used for bond strength tests showed that adhesive failure rather than cohesive failure occurred between any two layers. This implies that the original bond between a polymeric film (or aluminum foil) and the adhesive was a surface phenomenon and was strongly dependent upon the nature of the surfaces and the intimacy of contact between the surfaces. Cohesive failure implies that the original bond is cohesively bonded and that failure occurs within the adhesive rather than at the adhesive–film interface (7). Based on the type of failure that occurred between layers of the multi-layered materials, one can conclude that there was no interaction among the layers to enhance the mechanical properties of the multilayered materials.

Radiation Curing of Multilayered Materials. Considerable interest has been shown in the potential use of high-energy radiation to initiate polymerization or to modify polymers by processes such as crosslinking or degradation. When a polyolefin is exposed to ionizing radiation, the main effects are scission of main chains and creation of free radicals, cross-links, double bonds, and end-links. These chemical changes provide many interesting possibilities for polyolefin modification. Typical changes in physical and thermal properties resulting from the irradiation of an ethylene–butene copolymer are shown in Table III. Vicat softening and tensile strength at yield show small changes below 12 Mrad. Turner has reported that the crystalline melting point of polyethyene is depressed by 0.03–0.04°C/Mrad (8).

Table III.   Effect of Gamma Radiation on Properties of Ethylene–Butene Copolymer

Irradiation Dose[a] (Mrad)	Melt Index (g/10 min)	Vicat Softening Point (°C)	Tensile Strength at Yield (MPa)
0	0.3	121	28.3
1	0.1	121	28.2
2	0.5	121	28.3
6	<0.001	122	28.4
12	<0.001	125	29.0
18	<0.001	125	29.4

[a] Irradiated at 21–40°C.

Mechanical properties, such as elastic modulus and yield point, that depend on crystallinity per se are not seriously affected by low to moderate doses of ionizing radiation. On the other hand, those mechanical properties that are sensitive to interlamellar activity are most dramatically affected by the low to moderate radiation doses. This is seen in the ultimate tensile strength and elongation at failure of the polyolefins. It is also reflected in the large change in melt index between 0 and 18 Mrad, which indicates formation of cross-links that increase with increasing irradiation dose. The irradiated polyolefins have much higher toughness than the unirradiated polyolefins at temperatures of 70–120°C. In addition, the crosslinked polymer in this temperature range behaves like an elastomer and has significant strength above its crystalline melting point while the unirradiated polymer has no measurable strength and merely melts and flows.

Polymeric films that are used as the food-contacting layer in pouches must yield strong seals that remain fused during thermoprocessing. Table IV shows the effect of irradiation dose on seal strength of Pouches 1 and 2. Seal strength increases with increasing irradiation dose up to 6 Mrad. A significant reduction in seal strength was observed at 12 Mrad and more so at 18 Mrad. Both pouches yielded fusion seals up to 12 Mrad but tacky seals at 18 Mrad. Tacky seals are attributed to the irradiation-induced crosslinking in the ethylene–butene component of each pouch. In the case of Pouch 2, in which the food-contacting layer is a 70-30 blend of ethylene–butene copolymer and polyisobutylene, the seal strength of the pouch irradiated to 6 Mrad increased by 29% compared with a 43% increase for Pouch 1 that contained no polyisobutylene. The radiation-induced degradation of the polyisobutylene reduced the tensile strength and seal strength of the blend. The seal strengths of pouches 3 and 4 increased by 46% and 53%, respectively, after irradiation at 6 Mrad.

Table IV.   Effect of Radiation Dose on Pouch Sealability

Irradiation Dose[a] (Mrad)	Seal Strength (N/m)	
	Pouch 1	Pouch 2
0	245	255
1	274	265
2	304	274
6	353	323
12	323	314
18	78	98

[a] Irradiated at 21–40°C.

Bond strength data for four multilayered materials is shown in Table V. In each case the data is for the bond between the food-contacting layer and its adjacent layer. In Pouch 1, it is the bond between ethylene–butene copolymer and aluminum foil; in Pouch 2 between ethylene–butene copolymer—polyisobutylene blend and aluminum foil; in Pouch 3 between ethylene–butene copolymer and polyiminocaproyl; and in Pouch 4 between ethylene–butene copolymer and poly(ethylene terephthalate). Bond strength increased in the four multilayered materials after the irradiation treatment.

Table V.   Effect of Gamma Radiation on Bond Strength

Pouch Number	Bond Strength[a] (N/m)	Increase in Bond Strength After Irradiation[b] (%)
1	69	600
2	108	266
3	88	525
4	118	106

[a] Control.
[b] Dosage: 8 Mrad at 21–40°C.

Bond strength data for unirradiated and irradiated pouches (8 Mrad at 21–40°C) filled with beef slices, vacuum sealed, and retorted at 118°C for 40 minutes are shown in Table VI. In the case of the unirradiated pouches, Pouch 2 showed a 27% increase in the strength of the bond between the food-contacting layer and its adjacent layer. The other three unirradiated pouches showed a marked decrease in bond strength. Retorting of the four irradiated pouches caused no change in the strength of the bond between the food-contacting layer and the adjacent layer. With the unirradiated pouches, delamination occurred between the outside layer and the aluminum foil after the retorting; but no delamination occurred with the irradiated pouches.

Microscopic examination of specimens used on bond strength tests showed that adhesive failure occurred between the layers of the unirradiated materials and that cohesive failure occurred between the layers of the irradiated materials. Infrared spectroscopic analysis of test specimens showed evidence that the strong adhesion among layers was not caused by the mechanical interlocking of layers but by the formation of primary chemical bonds (intermolecular crosslinking) extending across the interface. For example, in the case of the ethylene–butene film-adhesive layer of Pouch 1, strong absorption bands arising from hydroxyl groups (10.9μ) in the unirradiated specimens were hardly discernible in the irradiated specimens while a new band attributed to the aliphatic ether group appeared at 9.6μ. Tensile testing offered addi-

Table VI.   Retortability of Pouches

Pouch Number	Bond Strength (N/m)			
	Unirradiated		Irradiated	
	Initial	Retorted	Initial	Retorted
1	69	39	490	490
2	108	137	382	490
3	88	59	529	529
4	118	49	245	245

tional evidence for the intermolecular crosslinking since the aluminum foil of the strongly bonded irradiated material underwent ductile plastic stretching without delamination or tearing away from the polymeric film. In the unirradiated material the aluminum foil tore and delaminated from the polymeric film because it contributes to the initial modulus of the material, but its load-bearing contribution falls off rapidly after about five percent elongation. In contrast, the aluminum foil carries the load all the way to failure in the irradiated material.

Conclusions

The strong adhesion between the aluminum foil and the food-contacting film is not caused by the mechanical interlocking aided by secondary valence forces but by the formation of primary chemical bonds extending across the interface between the aluminum foil and the adhesive and between the adhesive and the food-contacting film. The nature of the chemical reaction is not fully understood, but in each case the bond strengths of the irradiated multilayered materials indicate that a remarkably effective bond is formed. The irradiation-induced chemical changes in the multilayered materials were reflected in the superior performance of the retorted pouches of beef. Most important of all, there was no delamination among the layers of the irradiated pouch materials after thermoprocessing while delamination occurred among layers of the unirradiated pouch materials. Future work will center on the phenomenology of the peeling process and will elucidate the nature of the irradiation-induced chemical changes that occurred between layers of the multilayered materials.

Literature Cited

1. Duxbury, D. D., *Food Prod. Develop.* (1973) **8**, (7), 70.
2. Lampi, R. A., Rubinate, F. J., *Packag. Develop.* (1973) **3**, (3), 12.
3. Payne, G. O., Spiegl, C. J., Killoran, J. J., *Mod. Packag.* (1965) **38**, 106.
4. Holm, H. W., Jarrett, R. D., "Radiation Preservation of Foods," National Academy of Sciences–National Research Council, Publication **1273**, 1965.
5. Scop, P. M., Argon, A. S., *J. Compos. Mater.* (1967) **1**, 92.
6. Schrenk, W. J., Alfrey, T., *Polym. Eng. Sci.* (1969) **9**, (6), 393.
7. McKelvey, J. M., "Polymer Processing," p. 151, John Wiley & Sons, 1962.
8. Turner, D. T., Malliaris, A., Kusy, R. P., "Radiation Crosslinking of Polymers with Segregated Metallic Particles," Division of Isotopes Development, U.S. Atomic Energy Commission, Contract No.: **AT(30-1)-4110**, January 1972.

RECEIVED April 1, 1974.

Future Needs in Food Packaging Materials

SEYMOUR G. GILBERT

Department of Food Science and Packaging Science and Engineering Center, Rutgers University, New Brunswick, N.J. 08903

The future of food packaging and its material needs are examined in terms of societal changes. The expansion of packaged foods with increasing urbanization is imperative to cope with solid waste problems, but the future material choices may depend on their adaptability to ecologically effective systems. A cascade rather than recycle system is predicted because of health-related problems in controlling the safety as well as quality of recycled vs. virgin materials. Thus effective secondary and tertiary uses may be the most efficient system for conserving resources. The complexity of future societal demands will require rapid expansion of scientific knowledge which will fulfill these predictable future needs for packaging materials.

Food packaging reflects the needs of human societies and their technical capacities to provide for these needs. The kinds and amounts of food supply, the methods of food preservation, and the availability of packaging materials determine the packaging systems in any culture—be it prehistoric, present, or future.

In contrast to the agriculturally predominant past, the food packaging system in the highly industrialized modern society is based on a shift in the proportion of its food supply which originates at distant high productivity centers which are often closely associated with adjacent processing centers. Thus food in its final consumer form or as stabilized bulk commodities moves as a packaged product. The package is the major device for maintaining the stabilizing factors incorporated during preservation.

Some major new factors in modern packaging are the use of graphics for consumer appeal and additional design features for operational convenience. These features have often resulted in higher intrinsic package costs. However overall cost is reduced by greater sales volume. Lower production costs are usually obtained by higher capital expenditures for production machinery at the material fabrication and package production levels.

We expect that some of these factors will continue to influence strongly future food packaging. Certainly the availability of new materials will be a strong spur to the innovator. The interest in barrier polymers for plastic bottle production reflects the potential for these resins in the multibillion beverage containers market. The combination of physical strength with light weight had led to the rapid expansion of the plastic bottle market. Now, to penetrate this new market, high gas barrier properties are also needed for carbonated beverages.

How much the ecology question contributed to the impetus for developing plastic beverage bottles is debatable. I know from personal contacts this was originally a major factor. In part the uncertainty results from the need for clearer definition of the relation of packaging to environmental control.

The major contribution of packaging to solid waste control has often been ignored by less informed advocates. Food preservation, with its attendant packaging systems, has reduced biodegradable wastes associated with our food supply. This reduction applies not only to consumed food but to the wastes normally associated with non-edible portions. The separation of the production centers from the place of consumption has shifted the solid waste disposal problem to the production centers. At the consumer level a much smaller amount of packaging material has been substituted for waste disposal than was previously needed when primarily unprocessed food supplies were characteristic of urban life. The substitution of non-biodegradable packaging material for the much greater volume of biodegradable wastes has greatly aided in both health safety and in keeping the sewage loads from drowning our expanding cities. The relationship of packaging to safety in health and in the reduction of dangerous accumulation of sewage is most evident in hospitals and similar institutions such as nursing homes. The need for absolute control of microbial contamination in incoming medication and surgical devices is paramount. Prepackaged, sterile instruments are commonplace in the operating room, and the unit dose concept for medication is rapidly gaining acceptance.

What is less obvious but of greater potential volume is the growing use of prepackaged foods. Here the availability of safety and reliability by proper quality control helps to compensate for the skill of the individual cook, whose culinary expertise and sanitation are often impossible to control in a modern, large institution. The prospect of substantial cost reduction and the ability of prepackaged food to fit into schemes of federally subsidized geriatric care could be an important additional spur in this direction. The attendant reduction in solid waste from food preparation is again vital in this situation. Here, insect carriers of disease and their source of food, the garbage pile, require scrupulous control.

Beverages, in contrast to most foods, have a low ratio of nutrient to weight. Thus justification of packaging waste on nutrient supply is not possible. The industry could be a particularly vulnerable target for those concerned with the solid waste problem if this factor were the only basis for judgment. Established alternatives to the nonreturnable bottle in the noncarbonated areas are concentrates and soluble powders. The justification for moving water by container rather than pipe is made on the basis of pleasure rather than necessity where high quality municipal water is available. For the carbonated beverage industry to continue in a society concerned with urban solid waste, its packaging must be consistent with the available disposal systems.

Incineration, as well as returnables and recovered scrap, is an alternative to recycling. Because of the new polymeric barriers this technique can be used where the municipal incineration system incorporates energy recovery. This arrangement depends, however, on the availability of properly designed incinerators with scrubbers and energy recovery.

There is an important "feedback factor" which cannot be properly evaluated at this time. It concerns future municipal investment in a specific waste control system. This could result in legislation controlling the input of important potential waste materials to the municipality. For example large capital investment in a heat/energy recovery system based on incineration might induce legislative restrictions on low calorific materials like metals and glass.

Health-related institutions again show the future trend in a microcosm. Suitable incineration facilities may determine the ability of such institutions to function effectively in an urban environment. When adequate processes for solid waste and air pollution control by incineration are developed, the use of combustible, one-way trip disposables for all incoming materials can be expected to accelerate. This will result in a decrease in the need for a direct fuel supply.

The complexity of the problems associated with future food packaging can be seen from these examples. We have focused on only one factor in one segment of the food industry—the effect of change in solid waste disposal. The present uncertainties in technical and political aspects of the disposal problem prevent a definite decision as to whether packaging changes should be made. All the industry can do at present is prepare itself for a possible major shift that would have to be made in a short time.

Other factors for change are more certain of success. One is cost of the ultimate package. The steady rise in both real and inflated packaging costs causes higher retail food prices. However there is always consumer resistance to these increases. The packager will continue to search for alternate cheaper materials or those which can be fabricated more economically. The introduction of machines to replace hand labor or to operate at higher speeds may, as in the potato chip industry, lead to use of more costly materials which are more compatible with the newer machines. The key is always volume which in turn is related to total market and market share. If a well-capitalized company can cut overall costs by machine changes, it can consolidate its market position. Thus packaging changes leading to greater automation favor the larger companies. This is analogous to an autocatalytic reaction where the availability of initial capital is the equivalent of the limiting activation energy. In addition to the cost factor, changes which also result in more functional packages—e.g., longer shelf life, operational convenience, etc. —would encourage change.

Supply is related to cost. If basic materials become scarce, changes will follow. Thus constraints in the petroleum supply may affect the polymer industry, which is based almost entirely on petrochemicals. We can look for expanded use of paper-based materials to provide factors such as bulk, tensile properties, and thermal resistance combined with the low weights of proper barrier materials. Composites will continue to grow in quantity and complexity.

In addition to the solid waste problem, we can also expect that with expanding population or expanding demands of a static population, there will be societal pressure to reduce material usage over and above cost factors. These pressures could result in legislation to expand the environmental protection philosophy. For example, since plant materials are a renewable resource as well as readily recyclable, we may expect forced increases in paper-based packages.

The recycling concept has inherent constraints not usually recognized by its advocates. Since the difficulties of maintaining quality standards and control are often insurmountable in recycling heterogenous sources, it would be much more feasible to define a sequence of declining requirements for reused materials. This "cascade" principle would put food packaging at the first stage when stringent health safety considerations normally apply. Paperboard for food use would be restricted to either virgin pulp or its equivalent to avoid contamination problems such as the recent PCB (polychlorinated biphenyl) scare.

This idea is already embodied in the 1968 Food Amendments legislation in FDA. We expect the concept of close control of packaging material identity to be expanded with the legal and ethical ability to control the introduction of new packaging materials.

We can expect drastic changes in the food packaging industry. Some will arise from new consumer needs, new or expanded food supplies and products, and new food preservation systems such as aseptic packaging. Others will result from societally based constraints such as safety in health and hazardous use (as in the child safety closure legislation) and consumer protection against fraud, misinformation, or wrongful use where the burden for awareness is no longer on the buyer but on the supplier. Other such constraints will arise from environmental concerns or material scarcities.

The development of special societal needs, such as the large urban health center or geriatric institution with its special problems, will offer unique opportunities for specialized food packaging. The solutions to these problems may presage important priorities assigned in relation to broader aspects of a densely populated human society. A strong, scientifically based technology coupled with an attentive management will be needed to provide the ability to change rapidly if necessary. There must also be assurances against undue error but not at the price of unnecessary delay in meeting new challenges.

RECEIVED November 5, 1973.

INDEX

INDEX

A

Absorption in high nitrile
copolymers 69
Absorption of product by
polyethylene container 51, 58
Absorption values, dilute solution 70
Acid on applesauce stored in cans,
effect of ascorbic 38
Acrylonitrile–styrene copolymers .. 62
barrier properties of 65
vs. nitrile content, gas permeability of 67
vs. nitrile content, water permeability of 69
on permeability, effect of functional groups of 66
physical properties of 63
water and alcohol permeability of 68
Additives in cola beverage, taste/
odor detection of 72
Additives, packaging 16
Adhesion of end-sealing compounds 26
Alcohol permeability of acrylonitrile–styrene copolymers ... 68
Alcoholic beverages stored in
aluminum cans 43
Alkali extraction from container
soda–lime glass 17
Alloy of detinned tinplate,
iron–tin 31, 32
Alloy layer, tin–iron 2
Alloys, chemical composition limits
of wrought aluminum 36, 37
Aluminum
alloys, chemical composition
limits of wrought 36, 37
on canned beer, effect of 42
cans
alcoholic beverages stored in 43
aluminum pickup by a tea
beverage stored in 44
beer stored in 42
carbonated soft drinks stored in 43
fish stored in 41
frozen foods stored in 45
fruit and vegetable products
stored in 39, 40
meat products stored in 40
milk products stored in 41
tea beverages stored in 44
wines stored in 44
foil with foods, interaction of .. 45
-food compatibility test 37, 38
food packaging 45

Aluminum *(continued)*
for food packaging, compatibility
of 35
and health 46
packaging, flexible and semirigid 46
pickup by a tea beverage stored
in aluminum cans 44
on wines stored in aluminum
cans, effect of 43
Ammonia analysis for oxygen permeation through polyethylene,
copper– 54
Analysis, profile 73
Anti-pollution legislation 8
Applesauce stored in cans, effect of
ascorbic acid on 38
Aqueous extract 18
flint glass composition and 19
Aroma loss test results 57
Ascorbic acid on applesauce stored
in cans, effect of 38

B

Barrier polymers 96
Barrier properties of acrylonitrile–
styrene copolymers 65
Beef, thermal sterilization of
pouches of 89
Beer, canned 42
Beverages
high nitrile copolymers for
packaging 61
in polymeric containers, malt .. 65
stored in aluminum cans,
alcoholic 43
stored in aluminum cans, tea ... 44
taste/odor detection of additives
in cola 72
Biscuit mix packages 84
Bond strength, effect of gamma
radiation on 93
Bottle for cooking oil, high density
coated polyethylene 53
Bottle for flavor concentrates, high
density polyethylene 53
Bottle production, plastic 96
Breakage of flexible packaging,
mechanical 85
Butene copolymer, ethylene– 90

C

Can enamels, extractives from ... 31
Can enamels, flexibility of irradiated 26
Can-making materials 2
Can making, organic coatings in .. 6

Can-making technologies, new 9
Cans
  alcoholic beverages stored in
    aluminum 43
  aluminum pickup by a tea bev-
    erage stored in aluminum .. 44
  beer stored in aluminum 42
  carbonated soft drinks stored in
    aluminum 43
  cemented and welded 11
  drawn 10
    and ironed steel 10
  fish stored in aluminum 41
  for food, drawn 9
  frozen foods stored in aluminum 45
  fruit and vegetable products
    stored in aluminum 39, 40
  meat products stored in aluminum 40
  milk products stored in aluminum 41
  plain tinplate 2
  soldered tinplate 1
  tea beverages stored in aluminum 44
  tinplate 22
  wines stored in aluminum 44
Carbon dioxide permeability constants of commercial polymers 66
Carbonated soft drinks stored in
  aluminum cans 43
Cascade principle 98
Cast solder, microstructure of 30
Casting process, continuous 4
Cellophane, nitrocellulose-coated .. 80
Cellophane, poly(vinylidene chlo-
  ride)-coated 80
Cemented cans 11
Cheese packaging films 81
Chemical composition limits of
  wrought aluminum alloys 36, 37
Chemicals, low density polyethylene
  liner for various 52
Chemistry of base steel 4
Chromatography testing for per-
  meation loss, gas–liquid 55
Chromium-coated, tin free steel–
  electrolytic 4
Clean Air Act of 1970 8
Closures for glass containers 20
Coated cellophane nitrocellulose-.. 80
Coated cellophane, poly(vinyli-
  dene chloride)- 80
Coated polyethylene bottle for
  cooking oil 53
Coated polyethylene liner for cola
  concentrate 52
Coated on tinplate, enamels 23
Coating industry, container 8
Coating systems 8
Coatings
  on aluminum cans for fruit and
    vegetable products 40
  in can making, organic 6
  for glass 19
  organosol white 7
Code of Federal Regulations, FDA 16

Cohesive failure 91
Cola beverage, taste/odor detection
  of additives in 72
Cola beverage in Lopac and glass
  containers, profile of 74
Cola concentrate, polyethylene
  liner for 52
Color of applesauce stored in cans,
  effect of ascorbic acid on the 38
Commercial polymeric films for
  thermoprocessing 88
Compatibility
  of aluminum for food packaging 35
  flexible packaging and food
    product 77
  problems, package/product 82
  tests, aluminum–food 37, 38
  testing 78
Composition limits of wrought
  aluminum alloys, chemical ..36, 37
Container
  coating industry 8
  design, public health aspects of 12
  glass, soda-lime-silica 17
  and material criteria 63
  and product, interactions between
    polyethylene 51
  wall, absorption of product by
    the polyethylene 51, 58
  wall, contribution to the product
    from the polyethylene 58
Containers
  closures for glass 20
  extraction of 10 oz Lopac 71
  glass 15
  interaction of aluminum foil with
    foods in 45
  Lopac 61
  malt beverages in polymeric ... 65
  packaging food products in
    plastic 49
  for packaging irradiation-
    sterilized foods 22
  profile of cola drink in Lopac
    and glass 74
  sensory test for nitrile 71
  steam-sterilized food 8
  trends in the design of food ... 1
Copolymers
  absorption and permeation in
    high nitrile 69
  acrylonitrile–styrene 62
  barrier properties of acrylonitrile–
    styrene 65
  ethylene–butene 90
  for food packaging, high nitrile 61
  vs. nitrile content, gas permea-
    bility of acrylonitrile–styrene 67
  vs. nitrile content, water per-
    meability of acrylonitrile–
    styrene 69
  on permeability, effect of func-
    tional groups of acryloni-
    trile–styrene 66

Copolymers (continued)
   physical properties of acryloni-
     trile–styrene 63
   water and alcohol permeability
     of acrylonitrile–styrene ... 68
Concentrate, coated polyethylene
   liner for cola 52
Concentrates, polyethylene bottle
   for flavor 53
Construction of flexible film pouch 78
Continuous casting processes 4
Copper–ammonia analysis for
   oxygen permeation 54
Corrosion resistance 4, 6
   of enameled plate, undercutting 7
   of tinplate, effect of radiation on 30
Costs, packaging 98
Crack test for film, flex 82
Cracking of end-sealing compounds 26
Criteria affecting taste/odor of soft
   drinks 65
Criteria, container and material ...63, 65
Curing method, irradiation 87
Curing method, UV 9
Curing of multilayered materials,
   radiation 91

D

Design, food container 1, 12
Detection of additives in cola
   beverages 72
Detinned tinplate, iron–tin
   alloy of 31, 32
Dilute solution absorption values .. 70
Drawn cans 10
   for food 9
Drawn and ironed steel cans 10
Drink in Lopac and glass containers,
   profile of cola 74
Drink mixes in flexible packaging 84
Drinks, criteria affecting taste/odor
   of soft 65
Drinks stored in aluminum cans,
   carbonated soft 43

E

Electrolytic chromium-coated, tin
   free steel– 4
Electrolytic tinplate 2, 23
Enameled plate 7
Enamels 23, 25
   coated on tinplate 23
   extractives from 25, 31
   flexibility of irradiated can 26
End-sealing compounds 24, 25
   cracking and adhesion of 26
   rigidity of irradiated and
     unirradiated 27
Endurance tester, MIT flex 81
Energy recovery 97
Ethanol, analysis for 57
Ethylene–butene copolymer 90
   effect of gamma radiation on ... 91

External protective coatings for
   glass 19
Extract, aqueous 18
Extraction from container soda–lime
   glass, alkali 17
Extraction of 10 oz Lopac containers 71
Extractives from enamels 25, 31
Extractives after irradiation, change
   in amount of 33

F

Failure, cohesive 91
FDA-Code of Federal Regulations 16
Film pouch, construction of flexible 78
Films
   flex crack test for 82
   flex testing of cheese packaging 81
   flexible packaging 78, 79
   ketchup packaging 82
   mustard packaging 83
   polyethylene 80
   for thermoprocessing, commercial
     polymeric 88
   trace solvent removal in flexible
     packaging 84
Fish stored in aluminum cans ... 41
Flavor concentrates, polyethylene
   bottle for 53
Flex crack test for film 82
Flex endurance tester, MIT 81
Flex testing of cheese packaging
   films 81
Flexibility of irradiated can enamels 26
Flexible aluminum packaging 46
Flexible film pouch, construction of 78
Flexible materials, multilayered .. 88
Flexible packaging
   biscuit mixes in 84
   drink mixes in 84
   films, properties of 78, 79
   films, trace solvent removal in .. 84
   and food product compatibility.. 77
   ketchup in 82
   mechanical breakage of 85
   mustard in 83
   red meat in 83
Flexible pouches, multilayered ...87, 89
Flint glass composition and aqueous
   extract, relationship of 19
Foil, aluminum 45
Foil with foods, interaction of
   aluminum 45
Food
   compatibility testing,
     aluminum– 37, 38
   containers, design of 1
   containers, steam-sterilized 8
   drawn cans for 9
   packaging
     compatibility of aluminum for 35
     high nitrile copolymers for .. 61
     materials, future needs in ... 95
     specialized 99
     types of aluminum 45

Food (continued)
  product compatibility, flexible
      packaging and 77
  products in plastic containers .. 49
Foods, packaging thermoprocessed 87
Foods, prepackaged 96
Frozen foods stored in aluminum
      cans 44
Fruit products stored in aluminum
      cans 39, 40
Functional groups of acrylonitrile–
      styrene copolymers on perme-
      ability, effect of 66

G

Gamma radiation on bond strength,
      effect of 93
Gamma radiation on properties of
      ethylene–butene copolymer, ef-
      fect of 91
Gas–liquid chromatography testing
      for permeation loss 55
Gas permeability of acrylonitrile–
      styrene copolymers vs. nitrile
      content 67
Gas permeation of various polymers 68
Glass
  composition, flint 19
  containers 15
  closures for 20
  profile of cola drink in Lopac
      and 74
  external surface protective
      coatings for 19
  light transmission of soda–lime 20
  soda–lime–silica container 17
  test for alkali extraction from
      container soda–lime 17
Glassine 80
Grapes, volatile constituents of
      concord 56

H

Headspace analysis of polyfoil
      pouches 55
Health, aluminum and 47
Health aspects of container design,
      public 12

I

Incineration 97
Interactions between polyethylene
      container and product 51
Internal coatings on aluminum cans
      for fruit and vegetable prod-
      ucts 40
Irradiated can enamels 26
Irradiated end-sealing compounds 27
Irradiation
  change in amount of extractives
      after 33
  conditions 23
  for multilayered flexible
      pouches 89

Irradiation (continued)
  on cracking and adhesion of end-
      sealing compounds, effect of 26
  curing method 87
  of multilayered materials for
      packaging thermoprocessed
      foods 87
  -sterilized foods, tinplate con-
      tainers for packaging 22
Iron alloy of detinned tinplate,
      tin– 31, 32
Iron alloy layer, tin– 2
Iron on canned beer, effect of 42
Ironed steel cans, drawn and 10

J

Juices stored in aluminum cans .. 39

K

Ketchup in flexible packaging 82

L

Legislation, anti-pollution 8
Light transmission of soda–lime
      glass 20
Lime glass, light transmission of
      soda– 20
Lime glass, alkali extraction from
      container soda– 17
Lime–silica container glass, com-
      position of soda– 17
Limits of wrought aluminum alloys,
      chemical composition 36, 37
Liner for cola concentrate,
      polyethylene 52
Liner for various chemicals,
      polyethylene 52
Lopac containers 61
  extraction of 10 oz 71
  profile of cola drink in 74

M

Malt beverages in polymeric
      containers 65
Material criteria 63
Materials
  can-making 2
  future needs in food packaging 95
  multilayered flexible 88
  for packaging thermoprocessed
      foods 87
  radiation curing of multilayered 91
  supply of 98
Meat in flexible packaging 83
Meat products stored in aluminum
      cans 40
Mechanical breakage of flexible
      packaging 85
Migration influences, polymer 70
Milk products stored in aluminum
      cans 41
MIT flex endurance tester 81
Mix packages, biscuit 84
Mixes in flexible packaging, drink 84

Moisture on the gas permeation of
    various polymers, effect of ...   68
Multilayered flexible pouches   87
    irradiation conditions for   89
    testing methods for   89
Multilayered materials for packag-
    ing thermoprocessed foods ...   87
Multilayered materials, radiation
    curing of   91
Mustard in flexible packaging   83

N

Nitrile containers, sensory tests for   71
Nitrile content, gas permeability
    of acrylonitrile–styrene copoly-
    mers vs.   67
Nitrile content, water permeability
    of acrylonitrile–styrene copoly-
    mers vs.   69
Nitrile copolymers, absorption and
    permeation in high   69
Nitrile copolymers for food and
    beverage packaging, high   61
Nitrile polymers, dilute solution
    absorption values for   70
Nitrocellulose-coated cellophane ..   80

O

Odor, sources of   59
Oil, coated polyethylene bottle for
    cooking   53
Organic coatings in can making ..   6
Organosol white coatings   7
Oriented polypropylene   81
Oxygen permeability constants of
    commercial polymers   66
Oxygen permeation through
    polyethylene 50, 52
    copper–ammonia analysis for ...   54

P

Package criteria   65
Package/product compatibility
    problems   82
Packaging
    additives   16
    biscuit mixes in flexible   84
    compatibility of aluminum for
        food   35
    costs   98
    drink mixes in flexible   84
    films
        flex testing of cheese   81
        ketchup   82
        mustard   83
        properties of flexible 78, 79
        trace solvent removal in flexible   84
    flexible aluminum   46
    and food product compatibility,
        flexible   77
    food products in plastic containers   49
    high nitrile copolymers for food
        and beverage   61

Packaging (continued)
    materials, future needs in food ..   95
    mechanical breakage of flexible   85
    red meat in flexible   83
    to solid waste control, contribu-
        tion of   96
    specialized food   99
    thermoprocessed foods   87
    types of aluminum foil   45
Paper   80
Permeability
    of acrylonitrile–styrene copoly-
        mers vs. nitrile content, gas   67
    of acrylonitrile–styrene copoly-
        mers vs. nitrile content,
        water   69
    of acrylonitrile–styrene copoly-
        mers, water and alcohol ..   68
    constants of typical commercial
        polymers   66
Permeation
    of acrylonitrile–styrene copoly-
        mers, gas   67
    factors of polymer to meet
        package criteria   65
    in high nitrile copolymers   69
    loss, gas–liquid chromatography
        testing for   55
    through polyethylene, copper–
        ammonia analysis for oxygen   54
    through polyethylene, Permachor
        method to predict   50
    of various polymers, effect of
        moisture on the gas   68
Permachor method to predict per-
    meation through polyethylene   50
Pharmacopoeia standards, U.S. ..   17
Plastic bottle production   96
Plastic containers, packaging food
    products in   49
Plate, corrosion resistance of
    enameled   7
Pollution legislation, anti-   8
Polyethylene 50, 90
    bottle for cooking oil, coated ..   53
    bottle for flavor concentrates,
        high density   53
    container wall, absorption of
        product by the 51, 58
    container wall, contribution to
        the product from the   58
    copper–ammonia analysis for oxy-
        gen permeation through ..   54
    dilute solution absorption values
        for   70
    film   80
    liner for cola concentrate, coated   52
    liner for various chemicals   52
    oxygen permeation through   52
    Permachor method to predict
        permeation through   50
    and polyisobutylene, blend of ..   90
    product loss through 50, 55
    terephthalate   90

Polyfoil pouches, headspace
  analysis of 55
Polyisobutylene, blend of polyethylene and 90
Polymer migration influences 70
Polymeric containers, malt
  beverages in 65
Polymeric films for thermoprocessing 88
Polymers
  barrier 96
  dilute solution absorption values
    for 70
  the effect of moisture on the gas
    permeation of various 68
  to meet package criteria, required
    permeation factors of 65
  permeability constants of
    commercial 66
Polyolefins 91
Polypropylene, oriented 81
Poly(vinyl chloride) 70
Poly(vinylidene chloride)-coated
  cellophane 80
Pouches
  of beef, thermal sterilization of 89
  construction of flexible film 78
  headspace analysis of polyfoil .. 55
  irradiation conditions for multi-
    layered flexible 89
  multilayered flexible 87
  retorted 87, 93
  seal strength of 92
  testing methods for multilayered
    flexible 89
Prepackaged foods 96
Product
  compatibility, flexible packaging
    and food 77
  compatibility problems, package/ 82
  interactions between polyethylene
    container and 51
  loss through polyethylene 50, 55
  by the polyethylene container
    wall, absorption of 51, 58
  from the polyethylene container
    wall, contribution to the ...51, 58
Production, plastic bottle 96
Profile analysis 73
Profile of cola drink in Lopac and
  glass containers 74
Protective coatings for glass 19
Public health aspects of container
  design 12

Q

Quality as rolled, tin free steel– ... 5, 6
Quality, tin free steel-can maker's .. 5, 6

R

Radiation
  on bond strength, effect of gamma 93
  on corrosion resistance of tinplate,
    effect of 30

Radiation (continued)
  curing of multilayered materials 91
  dose on pouch seal strength,
    effect of 92
  on ethylene–butene copolymer,
    effect of gamma 91
  on tensile properties of solder,
    effect of 29
  on tensile properties of tinplate,
    effect of 28
Recycling 98
Retorted pouches 87, 93
Rigidity of irradiated and unirradiated end-sealing compounds .. 27

S

Seal strength, effect of radiation
  dose on pouch 92
Semirigid aluminum packaging ... 46
Sensory tests for nitrile containers 71
Shelf life testing 78
Shellfish stored in aluminum cans 41
Silica container glass, composition
  of soda–lime– 17
Soda–lime glass, light transmission
  of 20
Soda–lime glass, alkali extraction
  from container 17
Soda–lime–silica container glass,
  composition of 17
Soft drinks, criteria affecting
  taste/odor of 65
Soft drinks stored in aluminum
  cans, carbonated 43
Solder, tensile properties of 29
Solder, microstructure of cast 30
Solder, tinplate and 24, 28
Soldered tinplate cans 1
Solution absorption values, dilute 70
Solvent removal in flexible packag-
  ing films, trace 84
Specialized food packaging 99
Standards, U.S. Pharmacopoeia .. 17
Steam-sterilized food container ... 8
Steel
  –can maker's quality, tin free .. 5, 6
  cans, drawn and ironed 10
  chemistry of base 4
  cross section of tin-free 5
  –electrolytic chromium, tin free 4
  microstructure of MR-TU 29
  –quality as rolled, tin free 5, 6
Sterilization of pouches of beef,
  thermal 89
Sterilized food container, steam- .. 8
Sterilized foods, irradiation- 22
Styrene, copolymers of acrylonitrile
  and 62
Sulfur dioxide retention of wines
  stored in aluminum cans 43
Supply of materials 98
Surface protective coatings for glass 19

T

Taste/odor detection of additives
  on cola beverages 72
Taste/odor of soft drinks, criteria
  affecting 65
Taste test results, Triangle 73
Tea beverage stored in aluminum
  cans, aluminum pickup by a .. 44
Temperature on juices stored in
  aluminum cans, effect of 39
Tensile properties of solder 29
Tensile properties of tinplate 28
Test for alkali extraction from con-
  tainer soda–lime glass 17
Test for film, flex crack 82
Test procedures, aluminum–food
  compatibility 37
Test results, aluminum–food
  compatibility 38
Test results, aroma loss 57
Test results, Triangle taste 73
Test, Triangle difference 71
Tester, MIT flex endurance 81
Testing
  of cheese packaging films, Flex 81
  compatibility 78
  methods for multilayered flexible
    pouches 89
  for permeation loss, gas–liquid
    chromatography 55
  for nitrile containers, sensory .. 71
  shelf life 78
Thermal sterilization of pouches of
  beef 89
Thermoprocessed foods, multilayered materials for packaging .. 87
Thermoprocessing, commercial
  polymeric films for 88
Tin
  alloy of detinned tinplate, iron–.. 31, 32
  free steel–can maker's quality .. 5, 6
  free steel–electrolytic chromium
    coated 4
  free steel-quality as rolled 5, 6
  -iron alloy layer 2
Tinplate
  cans 22
  plain 2
  soldered 1

Tinplate *(continued)*
  containers for packaging irradia-
    tion-sterilized foods 22
  corrosion resistance of 30
  cross section of 552CR 3
  electrolytic 2, 23
  enamels coated on 23
  iron-tin alloy of detinned 31, 32
  and solder 24, 28
  tensile properties of 28
Trace solvent removal in flexible
  packaging films 84
Transmission of soda–lime glass,
  light 20
Triangle difference test 71
Triangle taste test results 73

U

Unirradiated end-sealing com-
  pounds 27
U.S. Pharmacopoeia standards ... 17
UV curing process 9

V

Vacuum of juices stored in aluminum cans, effect of temperatures on the 39
Vegetable products, internal coatings on aluminum cans for ...39, 40

W

Wall, absorption of product by the
  polyethylene container 51, 58
Wall, contribution to the product
  from the polyethylene con-
  tainer 58
Waste control, contribution of
  packages to solid 96
Water permeability of acrylonitrile–
  styrene copolymers 68
  vs. nitrile content 69
Water permeability constants of
  commercial polymers 66
Welded cans 11
White coatings, organosol 7
Wines stored in aluminum cans .. 44
  SO$_2$ retention of 43
Wrought aluminum alloys, chemical
  composition limits of 36, 37

The text of this book is set in 10 point Caledonia with two points of leading. The chapter numerals are set in 30 point Garamond; the chapter titles are set in 18 point Garamond Bold.

The book is printed offset on Danforth 550 Machine Blue White text, 50-pound. The cover is Joanna Book Binding blue linen.

Jacket design by Linda McKnight.
Editing and production by Virginia Orr.

The book was composed by the Mills-Frizell-Evans Co., Baltimore, Md., printed by The Maple Press Co., York, Pa., and bound by Complete Books Co., Philadelphia, Pa.